AKADEMIE DER WISSENSCHAFTEN DER DDR

Veröffentlichungen des Forschungsbereichs Geo- und Kosmoswissenschaften

Herausgegeben von Heinz Kautzleben

HEFT 15

Commission for Multilateral Cooperation of Academies of Sciences of Socialist Countries on the Complex Problem "Planetary Geophysical Studies"
CAPG-Project 3:
"Physical Properties of Mineral Systems under Thermodynamic Conditions of the Earth's Interior and Models of the Earth's Structure"

High Pressure Investigations in Geosciences

Edited by H. Stiller (GDR)

Associate Editors M. Kopf (GDR)
V. A. Kalinin, T. S. Lebedev (USSR)
M. Lastovickova, V. Kropacek (ČSSR)
F. Egerer (UVR), S. Franck (GDR)

With 126 Figures and 47 Tables

AKADEMIE-VERLAG BERLIN

1989

ISBN 3-05-500493-0
ISSN 0138-4600

Erschienen im Akademie-Verlag Berlin, Leipziger Straße 3—4, Berlin, DDR-1086

Lizenznummer: 202 · 100/445/88
Printed in the German Democratic Republic
Gesamtherstellung: VEB Druckhaus „Maxim Gorki“, Altenburg, 7400
Lektor: Dipl.-Met. Heide Deutscher
LSV 1405
Bestellnummer: 763 782 6 (2156/15)
04800

Contents

CHAPTER III

CHAPTER IV

CHAPTER V

CHAPTER VI

CHAPTER VII

Preface

This monograph is a kind of reflection of Project 3 "Physical Properties of the Mineral System under the Thermodynamic Conditions of the Earth's Interior and Models of the Earth's Structure" undertaken by the Commission for International Cooperation of the Academies of Sciences of the Socialist Countries in the complex problem "Planetary Geophysical Studies" (CAPG).

Experimental and theoretical papers dealing with the study of the physical properties of minerals and rocks under various temperature and pressure conditions are under consideration in this monograph. It also represents a summary of the results obtained within the past 5 years (1980—1985) by the Working Group Project 3. The results have become important for both, for the interpretation of geophysical measurements and observations and for the construction of theoretical models of the Earth's formation, evolution and present structure.

The monograph is the 4th report on investigations which have been performed since 1970 within Project 3. The former results were presented in:

- "High Pressure and Temperature Studies of the Physical Properties of Rocks and Minerals" — Naukova Dumka, Kiev 1977;
- "Theoretical and Experimental Investigations of Physical Properties of Rocks and Minerals under Extreme *p*, *T*-conditions" — Akademie-Verlag Berlin 1979;
- "Physical Properties of the Mineral System of the Earth's Interior" — Geophysical Institute, Prague 1985.

We should like to thank all the authors for their contributions and the co-editors for their help and cooperation.

We express our thanks to the publishing house and to H. Henschel and K. Weinreich for their help in preparing the manuscript for print.

H. Stiller
Vicepresident
Academy of Sciences
of the GDR

CHAPTER I

Elastic Properties

Spectral approach to determination of longitudinal elastic wave attenuation in rocks at pressures up to 1.5 GPA

G. A. Efimova, I. V. Zhukov, V. A. Kalinin

Attenuation of elastic waves is one of the most informative but also most hard-to-determine characteristics of rocks. The existing theoretical and experimental results of attenuation studies are contradictory [1, 2] which is attributed to the complexity of the objects under consideration to joint manifestation of different wave attenuation mechanisms, and to imperfections of experimental techniques. In the laboratory environment, high pressure contributes additional methodological difficulties that make it practically impossible to provide acceptable precision with the existing techniques for determination of elastic wave attenuation. For example, the difficulty of allowing for the losses on interfaces of the elements in a sonically tested system, on signal transformation and also on scaling and interference effect [5] is the disadvantage of the most common ultrasonic pulse echo-method that relies upon measuring amplitudes of multiple pulse reflexions from parallel faces of the sample [3, 4]. Moreover, the existing ultrasonic transducers, as a rule, have fixed resonance frequencies that define the maximal pulse amplitude measurable by the echo-method. This does not allow one to carry out measurements over a wide frequency range. Frequency modulation [2] enables determination of frequency characteristics by varying the basic pulse frequency, but in this case interference effects may also occur due to sample geometry. In the opinion of the present writers, spectral determination of wave attenuation as based on comparison of amplitudes of spectrum components and calculation of the attenuation constant of a given frequency by means of measuring spectrum amplitudes of incident and passing pulses [6] is an alternative approach to determination of attenuation in limited-size samples. This technique may be implemented rather easily for attenuation measurements under atmospheric pressure where ultrasonic transducers directly contact the faces of rock samples whose size practically is not limited, i.e. there is no need to take into account interference phenomena and the direct longitudinal wave under consideration may be readily discriminated against the background of reflected and converted waves. Contact losses in this case also would contribute significantly to the errors, but precision of attenuation determination may be improved by repeated measurements and accumulation of statistical data. Researchers experimenting with high-pressure setups, however, do not have such a possibility because much more time is required by each experiment. As applied to high-pressure studies, mostly to quasi-hydrostatic chambers, the direct use of the spectral method [6] is ineffective because of the following:

- it is difficult to take into account the pulse attenuation in the elements of acoustic section that have different frequency characteristics of wave attenuation;
- forms of recorded signal are more complicated due to the sample geometry limited by the design of high-pressure chambers;
- two receiving transducers are used, the transducer (at sample's "input") should be located in the high-pressure zone but must not be loaded over 500 MPa because of the fragility of piezoelectric ceramics.

The below proposed modification of the spectral method allows for the specific research environment of high quasi-hydrostatic pressures and requires no redesign of ultrasonic pulse installations. As the measuring system including both acoustic and electric sections is linear on the whole, i.e. output signal form is independent of input signal level [7], the spectral characteristic of the input signal is representable as

$$S(\omega) = \varrho(\omega)\, L(\omega)\, \varkappa K(\omega) \tag{1}$$

where $\varrho(\omega)$ is the input signal spectrum, $L(\omega)$ is amplitude-frequency characteristic of losses in the elements of acoustic and electric sections save for the sample, $\varkappa$ are losses in the interfaces between the sample and adjoining section elements (plastic envelope, in the case of quasi-hydrostatic setup), $K(\omega)$ is the desired wave attenuation frequency characteristic of the rock sample, attenuation coefficient $\alpha(\omega)$ being related with it by

$$\alpha(\omega) = -\frac{1}{l} \ln K(\omega) \tag{2}$$

where l is the sample length.

As may be seen from (1), in order to determine $K(\omega)$ one has to know the resulting characteristic $L(\omega)$ of overall system losses (except those taken into account by $\varkappa$), namely: losses for electro-acoustical signal transformation, losses in various acoustic contacts, losses in electrical system elements including the recording equipment. Studies of amplitude — frequency characteristics of each measuring system element involve numerous independent methodological problems. Therefore, compensation of $L(\omega)$ by introducing into the acoustic section a so called reference sample with known attenuation frequency characteristic and of the same size as the sample under consideration, seems to be optimal in this case. In doing so, output signal spectrum is representable as

$$S_e(\omega) = \varrho(\omega)\, L(\omega)\, \varkappa_e K_e(\omega) \tag{3}$$

where $K_e(\omega)$ is wave attenuation frequency characteristic of the reference sample that is known either from literature, or from experiments with massive samples under atmospheric pressure according to one of the existing techniques. Most suitable reference materials are metals (copper, brass, aluminium), because, as it was verified experimentally, attenuation of ultrasonic waves in them varies weakly under high hydro- and quasi-hydrostatic pressures. Thus, the rock attenuation coefficient may be determined from the ratio of characteristics (1) and (3):

$$\alpha(\omega) = \frac{1}{l}\left(\ln \frac{S_e(\omega)}{S(\omega)} - \ln \frac{\varkappa_e}{\varkappa} + \alpha_e(\omega)\, l_e\right). \tag{4}$$

Non-compensated contact losses, $\varkappa$ and $\varkappa_e$ are estimated by the formulas of coefficient of wave passage through the interface of two media with different acoustic resistances in the case of normal incidence:

$$\varkappa_e = \frac{4Z_0Z_e}{(Z_0 + Z_e)^2}; \quad \varkappa = \frac{4Z_0Z}{(Z_0 + Z)^2}; \quad Z = \varrho v_p; \quad Z_e = \varrho_e v_{pe} \tag{5}$$

where the speeds of elastic waves (longitudinal, in this case) v_p, v_{p_e} and the densities ϱ, ϱ_e of studied and reference samples, respectively, are determined experimentally in the course of sample immersion according to the technique [3]; and Z_0 is the acoustic resistance of the medium (plastic envelope) containing on the samples.

For the best compensation of diffraction divergence of the ultrasonic beam, it is recommendable to employ samples whose acoustic resistance is matched with that of the rock under consideration; since exact coincidence of Z and Z_e is unlikely, a difference in 20 to 25% between them would suffice because in this case correction for geometry is already negligible.

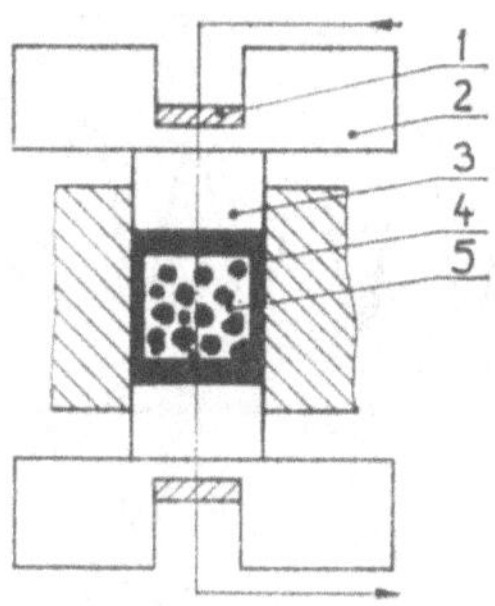

Fig. 1. Diagram of the acoustic set up of the measuring system.
(*1* — piezoelectric transducers; *2* — anvils; *3* — hard-alloy pistons; *4* — plastic envelope; *5* — sample).

Selection of the recorded pulse fragment to be subjected to frequency analysis is another methodological issue. The problem of seismic pulse spectrum distortions occurring at analysis of pulse fragment has been covered at length in seismology [8, 9]. In the experiments with quasi-hydrostatic pressure setups, the pulse fragment suitable for frequency analysis is defined by the thickness of elements of acoustic section schematically depicted in Fig. 1. The entrance of the first interference wave is, obviously, defined by the delay time of the wave reflected from the thinnest element. Owing to the lack of adhesive joints in the acoustic section of the high-pressure setup (stable contact was ensured by loading the system, and already at 0.1 GPa the recorded signal was stable in different sets of experiments), entrance time of the first interference wave and, consequently, the length of the fragment of nondistorted longitudinal direct wave depends on the thickness of quasi-hydrostatic envelope. In our experiments, the sides of lead envelopes were 1.5 to 2 mm high. Taking into account that the velocities of longitudinal waves in the lead lie between 2.1 to 2.4 km/sec and allowing for their variations due to pressure, pulse fragment of 0.5 mc/s length which is certainly free of interference was chosen for frequency analysis. In the general case, if the system involves thin layers or if pressure causes changes in envelope thickness that cannot be

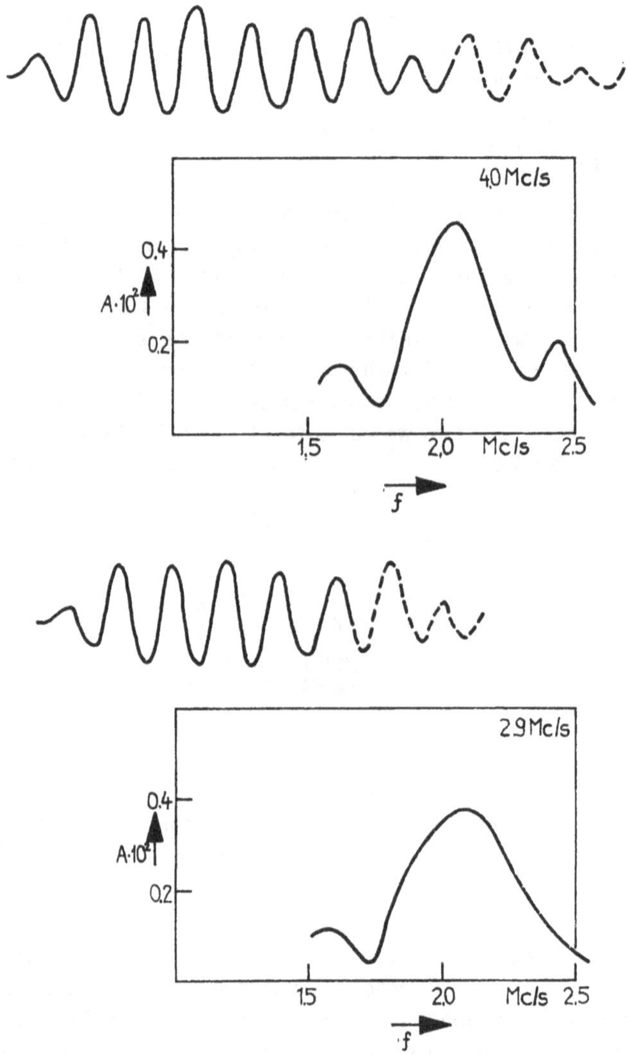

Fig. 2. **Forms** and spectra of recorded pulses vs. length of the analyzed fragment.

estimated, pulse distortion by interference of the direct and reflected waves is determined either visually by records, or by spectra that have in the case of interference, additional maximum vanishing with decrease of the length of analyzed fragment (Fig. 2). It should be noted that quantitative interpretation of the results of spectral analysis of incomplete pulse leads to errors that may be appreciable at spectrum sides [8]. Therefore, selection of the optimal record fragment for frequency analysis as applied to this particular case requires more careful substantiation. Comparison of the spectra of the studied and reference samples, all other system elements being unchanged, enables one to partially offset the error due to analysis incompleteness; in doing so, one has to compare spectra of fragments having equal number of half-cycles. In cases where it is impossible to compare pulses for rock sample and reference ones because of one or another process within the sample, it is very difficult to determine quantitative data via the proposed technique which greatly restricts its application domain. Solid-solid transition of calcite under 1.5 GPa exemplifies processes resulting in greater attenuation and dispersion of ultrasound speed. In this case, spectra may be compared only qualitatively [10].

Attenuation precision as determined by the above technique was estimated by calculating relative errors of determination of each term in (4). In particular, errors due to spectral analysis including digitalization of records were determined by means of comparing spectra of the same record fragment obtained at repeated analysis. Errors

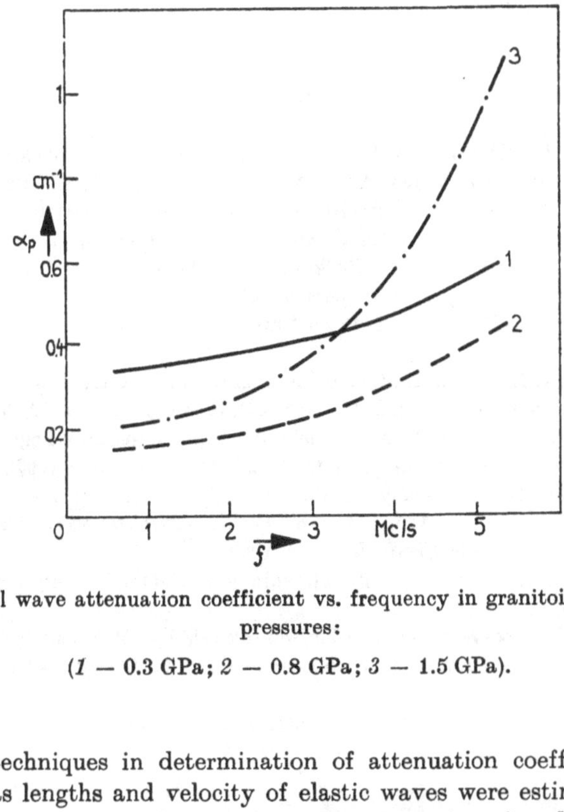

Fig. 3. Longitudinal wave attenuation coefficient vs. frequency in granitoid rock under various pressures:

(*1* — 0.3 GPa; *2* — 0.8 GPa; *3* — 1.5 GPa).

of the existing techniques in determination of attenuation coefficient in reference samples as well as lengths and velocity of elastic waves were estimated according to [3]. In our case, the resulting relative error in determination of α_p did not exceed 15%.

The above technique was tested on a granitoide rock, non-uniform fine-grained aplite. 20 mm high samples having 17.5 mm in diameter were loaded in the quasi-hydrostatic chamber up to 1.5 GPa. Brass was used as reference material its frequency characteristic of longitudinal wave attenuation was studied in [11]. The amplitude frequency analysis was carried out according to the program developed in the Department of Computing Geophysics at the Geophysical Institute. The longitudinal wave attenuation coefficient was determined for various pressures within the frequency range from 0.5 to 5.0 Mcps (Fig. 3). Notably, the calculated attenuation coefficients coincide with respect to the order of magnitude with data obtained for granitoids under atmospheric pressure. High pressure at first leads to the decrease of α_p at all frequencies under

0.8 GPa, and then to its increase under 1.5 GPa that is more pronounced at higher frequencies which is indicative of greater contribution of scattering on grain boundaries to the total attenuation of longitudinal waves under high pressures.

Conclusions. A technique has been developed for the determination of the elastic wave attenuation coefficient in the ultrasonic range for rocks under high quasi-hydrostatic pressures. It is built around the spectral technique for determination of attenuation. Calibration against reference sample is used for the sake of compensating frequency characteristics of measuring system elements. The error of the proposed technique is 15% at most which meets the objectives of studies of physical properties of rocks under high pressures.

References

[1] Volarovich, M. P.; Levykin, A. I.; Sizov, V. P.: Issledovanie zatukhaniya uprugikh voln v obraztsakh gornykh porod. Izv. AN USSR, ser. geofiz. (1960) 8, 1198–1203.

[2] Merkulova, V. M.: Chastothaya zavisimost' zatukhaniya ul'trazvuka v gornykh porodakh dlya megagertsevoj oblasti. Izv. AN USSR, Fiz. Zemli (1966) 8, 47–60.

[3] Volarovich, M. P.; Bayuk, E. I.: Fiziko-mekhanicheskie svoystva gornykh porod i mineralov pri vysokikh davleniyakh i temperaturakh. Nauka 1974.

[4] Truell, R.; Elbaum, Ch.; Chick, B.: Ultrasonic methods in solid state physics. N. Y., London 1969.

[5] Vanek, I.; Klima, K.; Pros, Z.: Osobennasti metodiki izmereniya pogloshcheniya uprugikh voln na obraztsakh gornykh porod. Izv. AN USSR, ser. geofiz. (1962) 5, 603–609.

[6] Merkulov, L. G.; Tokarev, V. A.: Fizicheskie osnovy spektral'nogo metoda izmereniya zatukhaniya ultrazvukovykh voln v materialakh. Defektoskopiya (1970) 4, 3–11.

[7] Alekseeva, V. G.: Raschet formy signalov. Energiya. Leningrad 1968.

[8] Berson, I. S.; Epinat'eva, A. M.: Dinamicheskie kharakteristiki sejsmicheskikh voln v real'nykh sredakh. Izv. AN USSR. Moscow. 1962.

[9] Molotova, L. V.: O dispersii obemnykh voln v terrigennych porodach. Izv. AN USSR, Fiz. Zemli (1966) 8, 36–46.

[10] Zhukov, I. V.: Nekotorye dinamicheskie kharakteristiki ultrazvukovykh protol'nykh voln pri fazovom perekhode kalitsita. v sb.: Aktual'nye problemy geofiziki, Nauka, Moscow (1985) 161–168.

[11] Papadakis, E. P.: Journ. Acoust. Soc. Amer. 37 (1965) 711.

A method of programmed modelling of thermobaric conditions of the lithosphere for studying mineral matter physical properties

T. S. LEBEDEV, V. A. KORCHIN

The degree of the still existing limitation of the recent experimental data on mineral matter physical characteristics at simultaneous *P-T* effect imitating the respective thermobaric conditions of different deep lithospheric zones is well known. E.g. the most reliable data on longitudinal wave velocity variation can only be obtained from a sufficient set of isobaric and isothermal experiments. The method of $V_p = f(H)$ calculation by isobar families, though universal, is somewhat defective. Its main defect is the difficult choice of their proper amount to obtain an optimal number of points for respective constructions without loosing some part of information, particularly in extreme areas. However, even in small sets of experiments this method of obtaining data on $V_p = f(H)$ changes is rather cumbersome and is practically not always justified. It is also easily seen that with insufficient isobaric and isothermal data on longitudinal wave velocity only its general depth dependence can be established [2, 8—11, 13, 17—28].

As an example, calculations of longitudinal wave velocity variation made by

$$\frac{dV_p}{dH} = \left(\frac{\partial V_p}{\partial P}\right)\frac{dP}{dH} + \left(\frac{\partial V_p}{\partial T}\right)\frac{dT}{dH}$$

for three different temperature-depth functions may be used [15].

From the calculated curves shown in Fig. 1 it follows that $V_p = f(H)$ changes depend on *P-T* conditions of different depths, more precisely on their $V_p = f(P)$ and $V_p = f(T)$ gradients.

Analyses of the calculations based on corresponding laboratory $V_p = f(P)$ and $V_p = f(T)$ determinations made one at room temperature and the other at atmospheric pressure evidence their incapability of giving a high confidence since the former and the latter case disregard the destructive action of temperature [9, 10] and the compensating effect of pressure on the destruction by heating [11], respectively. These phenomena evidently occur at great depths.

The experiments show that the constant compensating pressure decreases significantly the temperature dependence of the velocity variation gradient [9, 10, 13, 21—24, 30]. If the $V_p = f(H)$ calculations regard one of $V_p = f(T)$ dependences at some $P = \text{const} > 0.1$ MPa, the velocity functions slightly decreasing initially at a definite depth (Fig. 1, curves 3, 6) will increase again. This yields two extreme $V_p = f(H)$ zones ($V_{p_{max}}$ and

$V_{p_{min}}$ occuring possibly at depths of 2—8 km and 10—15 km, respectively). Fig. 1 also shows their depths to depend significantly on the assumed temperature gradient.

Unfortunately, such $V_p = f(H)$ calculation methods characterize the velocity-depth function only qualitatively since the calculations are based on the $V_p = f(P)$ and $V_p = f(T)$ data obtained at room temperature and a certain compensating pressure respectively. Thus, the most real velocity distribution at depths can only be established by a sufficient amount of isotherms and isobars, each of them giving one V_p value corresponding with combined P-T for the region (depth) under study.

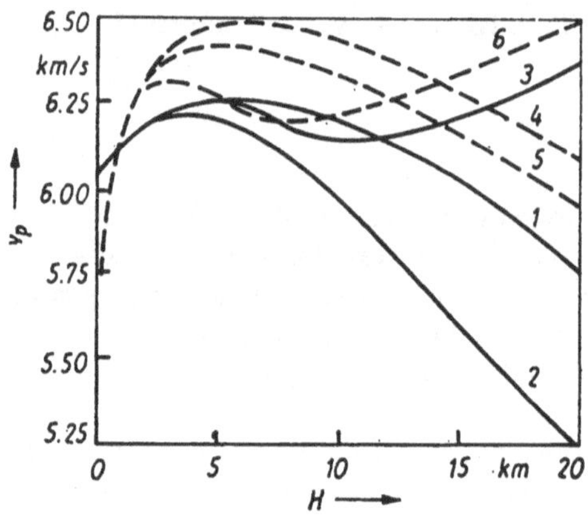

Fig. 1. Calculated V_p changes with depth for two different rocks. Curves: *1, 4* — dT/dH (min) and *2, 5* — dT/dH (max) at $P = 0.1 \cdot 10^{-3}$ GPa; *3, 6* — dT/dH (mid) at $P = 0.5$ GPa. Solid and dashed lines refer to first and second rock respectively.

Considering the above facts and other circumstances we propose a new method of programmed modelling in high pressure chambers with samples affected simultaneously by P and T imitating different deep P-T conditions of concrete geologic objects or entire regions [5, 8, 11, 12, 17, 19, 25—29].

Assuming that pressure of deep crustal zones is hydrostatic and depends only on mineral matter density and gravity distribution, the pressure at a depth H can be calculated by

$$P_{(H)} = 4\pi g \int_h^H \frac{\varrho}{h^2} \, \mathrm{d}h \int_0^h \varrho x^2 \, \mathrm{d}x,$$

where P, g, h, x, and ϱ are hydrostatic pressure, gravity constant, integration variables and depth-dependent mineral matter density, respectively.

Supposing the Earth's crust of a concrete region to consist of n layers each having a constant density ϱ_n and thickness h the above expression can be simplified as follows:

$$P_{(H)} = g_0 \sum_n \varrho_n h_n,$$

where g_0 is the gravity value at the day surface. Note also a small correction for P due to ϱ_n change with depth which can be calculated graphically from the experimental data.

Many researchers assume the average density of the upper (to 15 km) lithosphere mineral matter to be 2.65—2.77 and that of deeper lithosphere and upper mantle to be 2.78—2.90 and $3.27-3.52 \cdot 10^3$ kg m^{-3}, respectively. The experimental data on specific volume deformation and compressibility of rocks suggests a 3—4% density increase for acid but only 2.4—3.2% one [1, 16] for basic and ultrabasic rocks with pressure increase from atmospheric to that observed at depths of ca. 80 km.

Most of the Ukraine regions exhibit a mean density value of $2.71 \cdot 10^3$ kg m^{-3} for upper crust acid rocks increasing to $2.73 \cdot 10^3$ kg m^{-3} at depths of 15—20 km, that of basic ones being ca. $3.00 \cdot 10^3$ kg m^{-3} below 10 km.

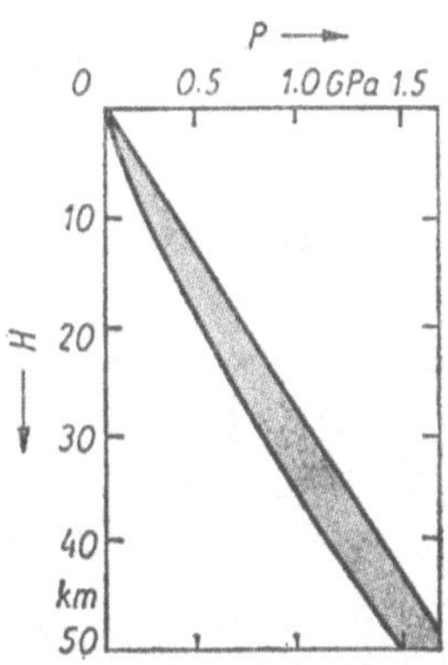

Fig. 2. Range of possible hydrostatic pressure variations with depth for different parts of USh.

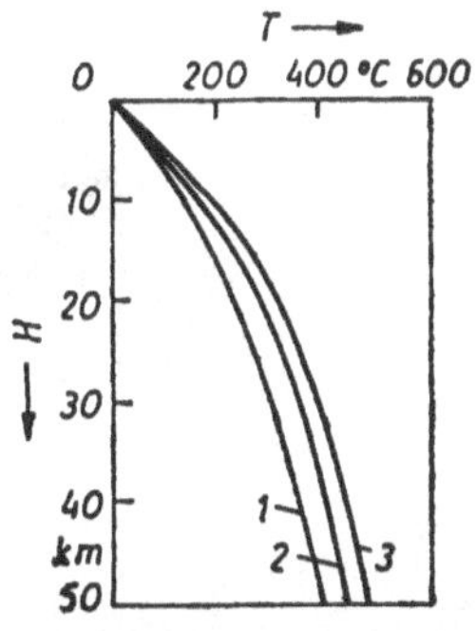

Fig. 3. T-variations with depth for different parths of USh. *1* — southwestern and northwestern, *2* — central, *3* — southeastern and Near-Azov part.

The mineral matter density and supposed deep structure data for definite crustal areas (as obtained by explosion seismology and other geophysical methods) as well as the assumed relative rock homogeneity within the layers enable us to calculate the hydrostatic pressure for the respective depth (Fig. 2). In the Ukrainian Shield, the hydrostatic pressure varies from atmospheric to 1.5 GPa at depths from the day surface to 50 km, its mean scatter for different shield parts being $\pm 2\%$ at the greatest depths.

The temperature variation with depth (Fig. 3) was calculated by geothermal data [6, 14]. The calculation error reaches tens of degrees even for areas with a practically stationary heat flow.

Thus, the data on depth dependence of pressure and temperature (Fig. 2, 3) enable us to set up programs of P-T variations in chambers of high pressure equipments to imitate experimentally possible deep thermobaric conditions of concrete geologic regions. E.g. for the Ukrainian Shield three programs of P-T regimes of our experiments (Fig. 4) have been calculated.

The experimental thermobaric programs were prepared for the hydrostatic conditions of the pressure distribution in the lithosphere though possible regional inhomogeneities of the whole vertical stress component of mineral masses at different depths are inferred from its multiform tectonic activity [4]. The laboratory experiments combined with tectonophysical and earthquake energy data as well as mine data and those of an analysis of tectonic motion velocity gradients illustrate [4] the reality of maximum tangential stresses in the pressure range from 10 ± 5 (tectonically calm regions) to 100 ± 50 MPa (active zones). Thus, for the Ukrainian Shield the possible deviation from hydrostatic pressure can be $\pm 1-2\%$ of the maximum one for the respective depths. Also known for the Ukrainian Shield is an ununiform temperature distribution at depths (Fig. 3), the irregular horizontal temperature gradient and ununiform rock physical properties distribution in the upper lithosphere causing thermoelastic stress and its concentration in some zones [14]. The evaluation of the thermoelastic stress magnitude order [7] shows that stress may cause deviation in pressure distribution to $\pm 5-10\%$ at different depths in the upper lithosphere of the Ukrainian Shield. Thus, in the interior of the Ukrainian Shield pressure may deviate by $\pm 3-12\%$ from the hydrostatic one.

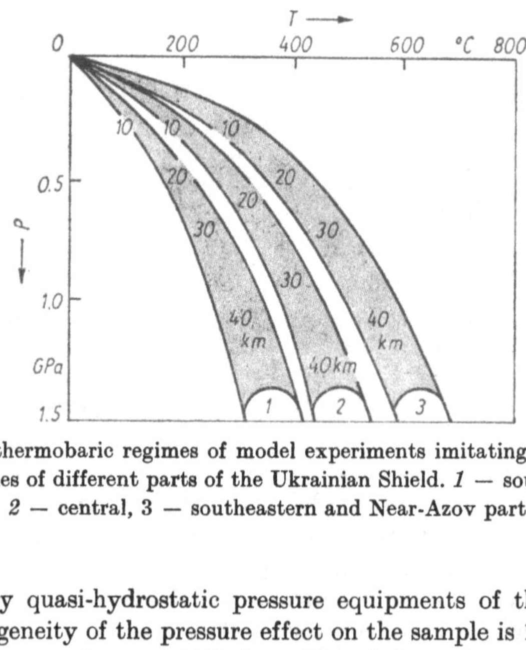

Fig. 4. Programs of thermobaric regimes of model experiments imitating *P-T* conditions of the upper lithosphere zones of different parts of the Ukrainian Shield. *1* — south- and northwestern; *2* — central, 3 — southeastern and Near-Azov part.

In the laboratory quasi-hydrostatic pressure equipments of the "cylinder-piston" system the inhomogeneity of the pressure effect on the sample is $1-2\%$ by calculated [3] and $5-10\%$ by experimental [13] data. This defines the quasi-hydrostatic high *P-T* experiments in which *P-T* change according to the programs shown in Fig. 4 as imitating one of the possible *P-T* conditions of the lithosphere of such a presently calm tectonic region as the Ukrainian Shield (USh).

In elaborating the method of programmed modelling its similarity to that defining the velocity-depth function by isobar families (Fig. 5, 6) was observed (most deviations did not exceed $1-1.5\%$). This qualifies our new method of programmed modelling aimed at studying V_p, $V_s = f(H)$ as more simple and more informative.

In what follows we discuss briefly the practical features of the programmed *P-T* experiments made with high hydrostatic and quasihydrostatic pressure equipments. The necessary *P-T* conditions in the chambers are most easily created by a continuous programmed *P* increase (Fig. 4) with a slow temperature elevation. In each concrete case, the realization of the changes in programmed *P-T* combinations in the chamber of the high *P* equipment depends on its construction, heater characteristics and methods

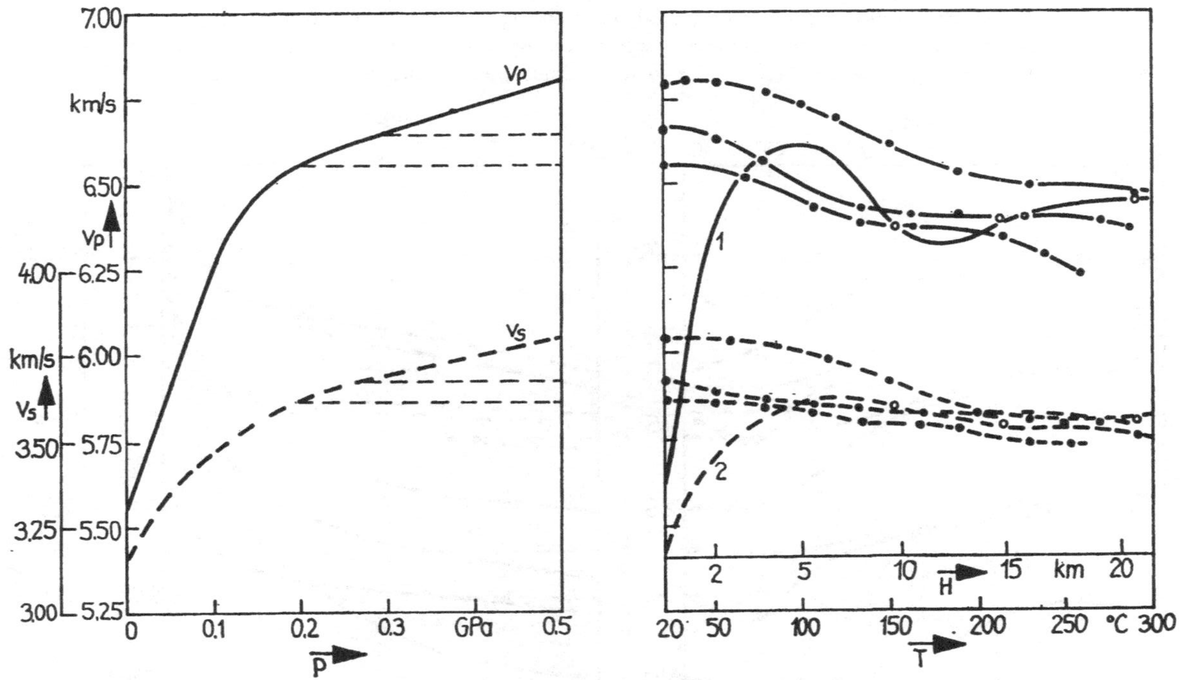

Fig. 5. Character of V_p and V_s changes in a diorite sample depending on pressure (a) and temperature (b) at various constant pressure. *1, 2* — $V_p = f(H)$, $V_s = f(H)$ — programmed modelling method. Solid, dashed lines and circles indicate V_p, V_s-values and coincidence points for V_p and V_s-values in programmed and isobaric regimes, respectively.

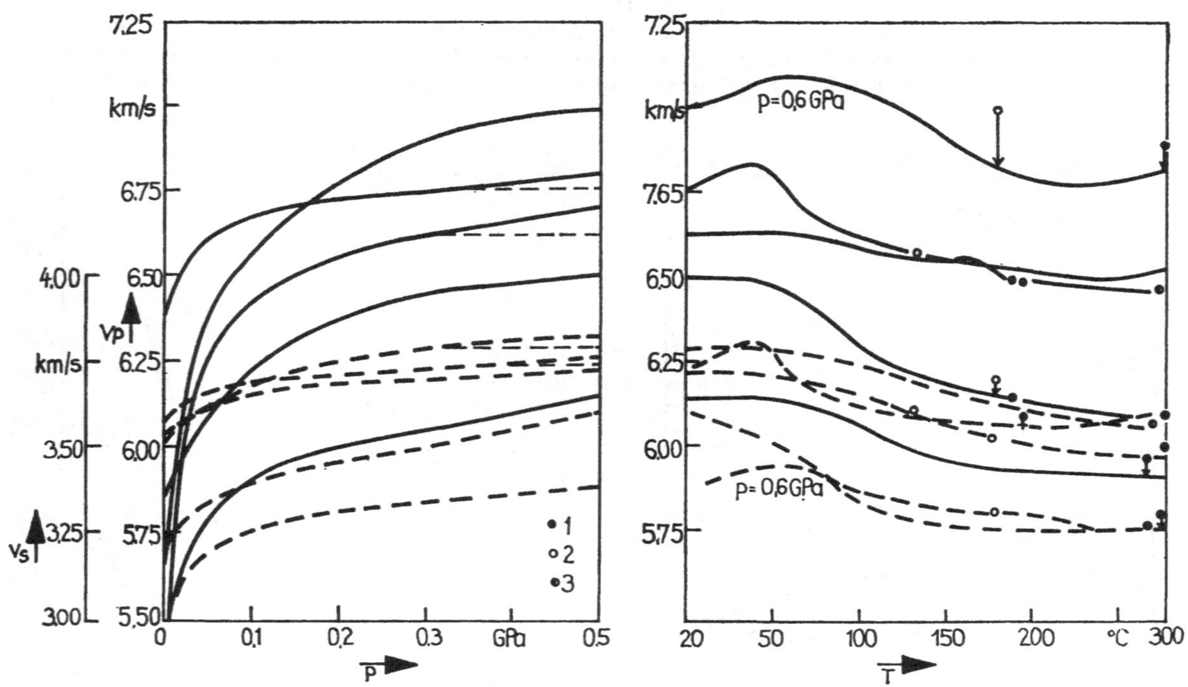

Fig. 6. $V_p = f(P)$ and $V_s = f(P)$ curves (a) and comparison of $V_p = f(PT)$, $V_s = f(PT)$ data (b) as obtained by programmed modelling method (1, 2, 3) and in isobaric regime for some rocks. *1, 2, 3* — are various PT programs of experiments. Solid and dashed lines indicate V_p and V_s, respectively.

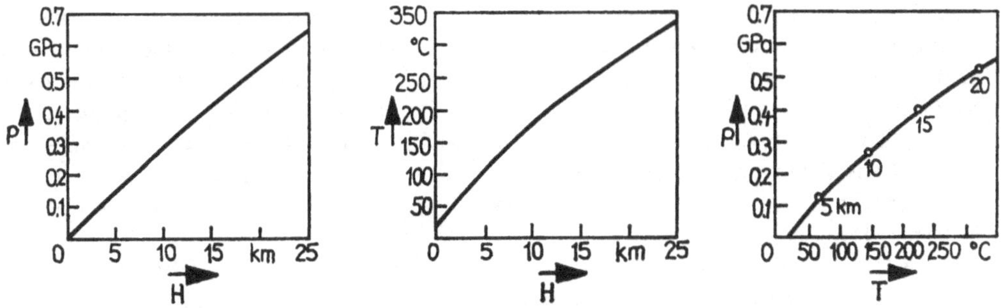

Fig. 7. a and b — possible changes of P and T with depth, c — principal scheme of programmed P-T experiments.

of P-T measurement. E.g. if the heater operation time (beginning from its "on" instant) corresponds well with the measured sample temperature, the experimental programs can tie to the experiment duration. This feature is used in V_p and V_s studies when samples are effected by P and T simultaneously.

For each heater linear or quadratic dependence of sample temperature on the experiment duration can be obtained in the operational regime chosen (under the control of tension U and intensity I got by the heater). This dependence defines the prescribed continuous or stepwise operation regime of the piston device of the press, i.e. two modes of the programmed experiments. The dependence of the sample temperature variation time on U and I of the heater enables the user to avoid frequent T measurements in the experiment except for general control ones. In most experiments the physical parameters were measured at intervals of 5 min.

We have considered methodical characteristics of programmed P-T experiments made by high quasi-hydrostatic pressure and temperature apparatus. In preparing an

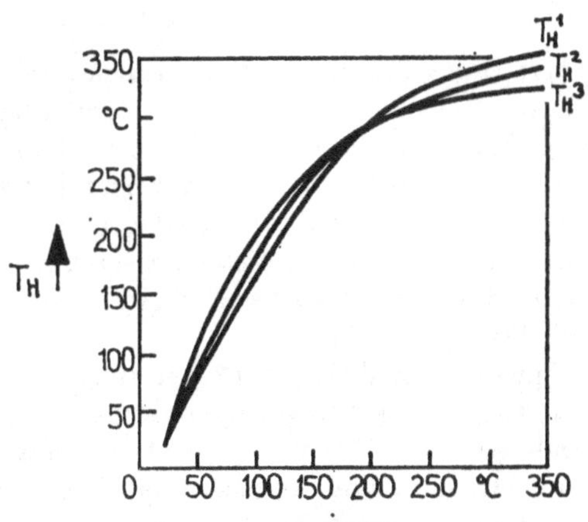

Fig. 8. Calibrating diagrams for sample temperature determinations in the chamber (T_c) by T measurements made in different parts of heating element of the high quasi-hydrostatic pressure apparatus (T_h^1, T_h^2, T_h^3).

experiment the coefficients of recalculation of press forces into quasi-confining pressure acting on the sample in the chamber at a respective temperature the time dependence of its variation should be known beforehand. The P-T depth function obtained is used to prepare tables and diagrams (e.g. Fig. 7) illustrating the P-T change program chosen. Then the pressure values are recalculated into press force units ($F = SP + \beta$, where S and β are plunger piston cross-section area and correction for friction in the chamber respectively) and the temperature values into operation time intervals of the heater (t) (Fig. 8, 9).

Since the temperature in high pressure apparatus begins to increase from the moment of switching on the heater in a strictly defined regime time intervals of F-changes required for each of experimental programs accepted can be easily calculated, i.e. $F = f(t)$ will agree with $P = f(T)$ in the experiment.

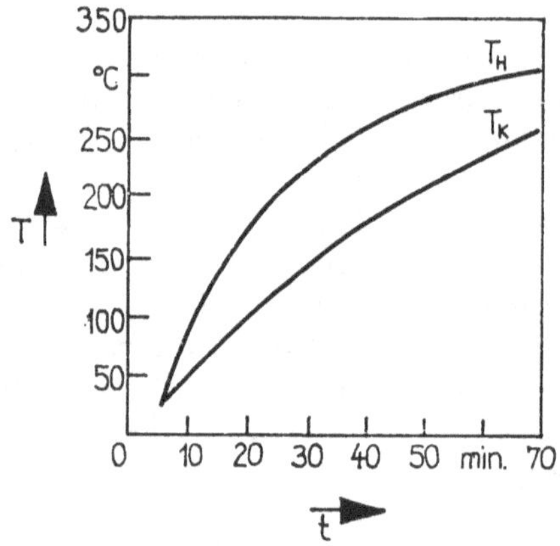

Fig. 9. Temperature measurement at one of heating element points (T_h) in the chamber of the high quasi-hydrostatic pressure apparatus (T_c) depending on heating time (t).

The first method of performing programmed experiments is used in the case where modelled P-T conditions enable us to choose dT/dt (by regulating the heater operation) and dP/dt (by stabilizing this value with a special device at the press) in such a way that pressure and temperature changing in a semiautomatic continuous regime in the chamber will correspond with prescribed deep P-T conditions of a definite region (Fig. 7). The mineral matter physical characteristic measurement and the piston motion registration with dial indicators were made each 5 min., one complete experimental cycle lasting 60—70 min (Fig. 10a).

The second discrete-stepwise method (Fig. 10b) investigates samples in one or several modes of P and T distribution. In such cases the chamber temperature also increases depending on the predefined conditions of the heater operation whereas the press pressure elevates after some time intervals. E.g. shortly after physical parameter determination at P_1 and T_1, a higher temperature T_2 is used in the measurement made at the same pressure P_1. Then the pressure increases to P_2, new measurements are made at T_2 and so on (Fig. 10b). This method, more precisely programmed modelling modification, provides necessary information on one and the same sample simultaneously in two or more P-T relations.

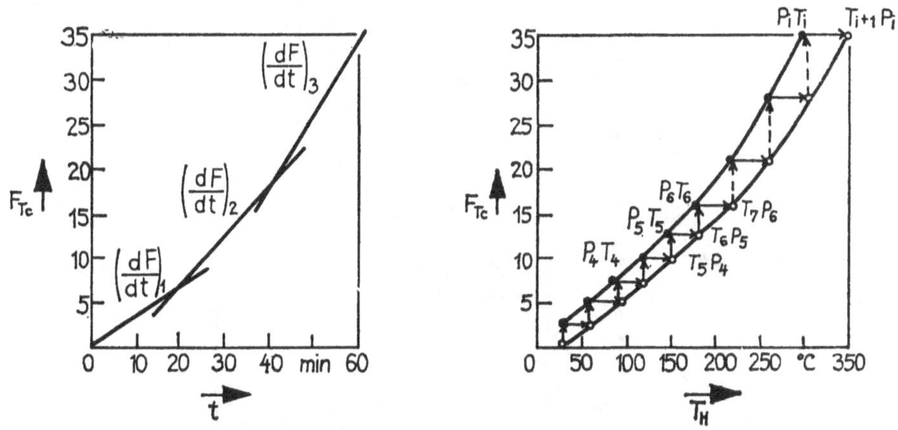

Fig. 10. Methods of changes of press forces (F) in programmed experiments. a — continuous semiautomatic method with two corrections, b — discrete-stepwise methods (for two simultaneous programs).

When the elastic properties of rocks are studied within a region with somewhat uncertain P-T-depth functions some changes in the experiment run occur. The velocities are studied in the isothermic and isobaric regimes at short P-T intervals (Fig. 11). The highest and lowest P values at respective $T_{\min}$ and $T_{\max}$ indicate possible deep regional P-T conditions. Such a program provides two or three specific dependences of V_p, $V_s = f(H)$.

For crustal parts differing significantly as to deep P-T conditions experimental programs may be used of which one is exemplified by Fig. 12. Their typical feature is that first isothermal determinations are made at T slightly exceeding the room tempe-

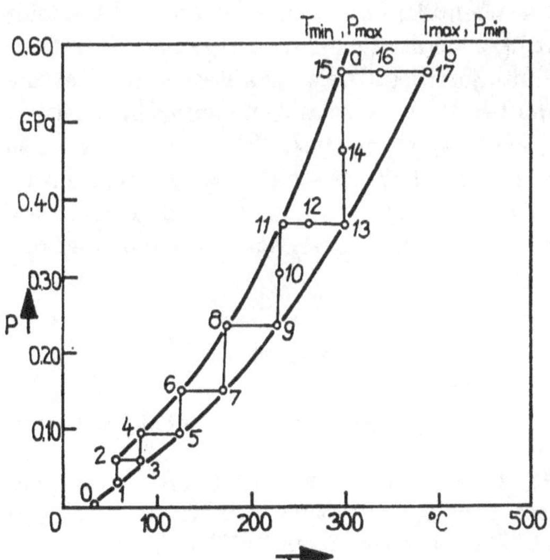

Fig. 11. Experimental program for two possible experimental deep P-T regimes. a — $T_{\min}$, $P_{\max}$; b — $T_{\max}$, $P_{\min}$. The numbers indicate V_p and V_s measurement points.

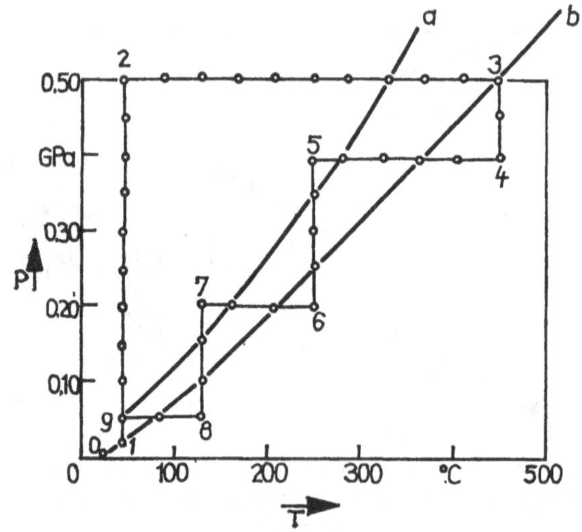

Fig. 12. Program of measurement of four isotherms (1—9—2, 3—4, 5—6, 7—8) and four isobars (2—3, 4—5, 6—7, 8—9) in one experiment. a, b — assumed deep *P-T* conditions.

rature and P ranging to P_{max} that depends on *P-T*-equipment abilities and experimental tasks. Then isobaric measurements are made to T_{max} at P maximally approaching P_{max}. Afterwards, isothermal and isobaric velocity measurements are made cooling the chamber at definite P and T intervals respectively. Fig. 12 shows a program providing four isobars and four isotherms, which yields points to set up curves V_p, $V_s = f(H)$. Note that programs given in Fig. 11 and 12 are only realized on condition of accurate P and T determinations in the chamber.

Thus, our new experimental method of programmed modelling of a simultaneous effect of *P-T* simulating concrete (general or regional) deep thermobaric regimes may be concluded to be quite justified. It has enabled detailed studies of physical characteristics of a great number of eruptive and metamorphic rock samples under different *P-T* conditions in a relatively short time, the general task of the model *P-T* experiments being to provide essentially new valuable information on peculiar changes of mineral matter physical parameters under *P-T* conditions, which could maximally approach real *P-T* situations of the interior at the present stage of experimental development.

References

[1] BELOUSOV, V. V.: Earth's crust and upper mantle of the continents (in Russian). Moscow, Nauka (1966) 123 p.

[2] VOLAROVICH, M. P.; BAYUK, E. I.; SHAGINYAN, G. SH.: Investigation of longitudinal wave velocity and absorption in minerals at pressures to 15 kb and temperatures to 600 °C (in Russian). Izv. AN SSSR, Ser. Fizika Zemli (1977) 7, 82—90.

[3] VOLAROVICH, M. P.; BAYUK, E. I.; LEVIKIN, A. I. et al.: Physical and mechanical properties of rocks and minerals at high pressures and temperatures (in Russian). Moscow, Nauka (1974) 223 p.

[4] GZOVSKY, M. V.: Principles of tectonics (in Russian). Moscow, Nedra (1975) 536 p.

[5] Studies of physical properties of Earth's mineral matter at high thermodynamic parameters (in Russian). Kiev, Naukova dumka (1977) 220 p.

[6] KUTAS, R. I.; GORDIENKO, V. V.: Thermal field of Ukraine (in Russian). Kiev, Naukova dumka (1971) 139 p.

[7] LANDAU, L. D.; LIFSHITS, E. M.: Mechanics of solid media (in Russian). Moscow, Vysshaia shkola (1963) 371 p.

[8] LEBEDEV, T. S.: Physical properties of lithospheric rocks at great depths. In: Problems of Earth's physics in Ukraine (in Russian). Kiev, Naukova dumka (1975) 9–10.

[9] LEBEDEV, T. S.; SHAPOVAL, V. I.; KORCHIN, V. A.: Investigations of high P and T effect on velocity variation in eruptive and metamorphic rocks (in Russian). Geofiz. sborn., Kiev (1969) 31, 75–93.

[10] LEBEDEV, T. S.; SHAPOVAL, V. I.; KORCHIN, V. A.: Studies of high P-T effect on elastic wave velocities in some rocks of the Ukrainian Shield (in Russian). Geofiz. sborn., Kiev (1969) 32, 5–17.

[11] LEBEDEV, T. S.; SHAPOVAL, V. I.; KORCHIN, V. A.: New data on longitudinal wave velocity in rocks at high thermodynamic parameters (in Russian). Geofiz. zborn., Kiev (1972) 49, 9–27.

[12] LEBEDEV, T. S.; SHAPOVAL, V. I.; KORCHIN, V. A.: Programmed modelling of deep thermodynamic conditions in experimental studies of mineral matter elastic properties. In: Geophysical studies of the Earth's crust and upper mantle of Ukraine (in Russian). Kiev, Naukova dumka (1975) 25–30.

[13] LEBEDEV, T. S.; KORNIETS, D. V.; SHAPOVAL, V. I. et al.: Elastic properties of rocks at high pressures (in Russian). Kiev, Naukova dumka (1972) 183 p.

[14] LUBIMOVA, E. A.: Thermics of Earth and Moon (in Russian). Moscow, Nauka (1968) 279 p.

[15] MAGNICKI, V. A.: Inner structure and physics of the Earth (in Russian). Moscow, Nedra (1965) 379 p.

[16] Handbook of physical constants of rocks (in Russian). Moscow, Mir (1969) 542 p.

[17] Physical properties of rocks and minerals at high pressures and temperatures (in Russian). Proc. IVth All-Union Meet., Tbilisi, Metsniereba (1974) 381 p.

[18] Physical properties of rocks at high pressures and temperatures and problems of seismology (in Russian). Proc. VIth All-Union Meet., Tashkent, Fan (1981) 200 p.

[19] Physical properties of rocks at high thermodynamic parameters (in Russian). Proc. IIId All-Union Meet., Kiev, Naukova dumka (1971) 284 p.

[20] Physical properties of rocks at high thermodynamic parameters (in Russian). Proc. Vth All-Union Meet., Baku, ELM (1978) 308 p.

[21] HUGHES, D. S.; GROSS, J. H.: Elastic wave velocities in rocks at high pressure and temperature. Geophysics **16** (1951) 4, 577–593.

[22] HUGHES, D. S.; JONES, H. J.: Variation of elastic moduli of igneous rocks with pressures and temperatures. Bull. Geol. Soc. Amer. **61** (1950) 843–856.

[23] HUGHES, D. S.; MAURETTE, C.: Variation of elastic wave velocities in granites with pressure and temperatures. Geophysics **21** (1956) 2, 277–287.

[24] HUGHES, D. S.; MAURETTE, C.: Determination des vitesses d'onde elastique dans diverses roches en fonction de la pression et de la temperature. Rev. Inst. Franc. Petrole et Ann. combustible, liquide (1957) 12, 730–752.

[25] LEBEDEV, T. S.: Investigations of some physical properties of mineral matter under thermodynamic conditions of the Earth's interior. Zeitschrift für Geologische Wissenschaften, Berlin **3** (1975) 6, 717–739.

[26] LEBEDEV, T. S.: Elastic properties of rocks under experimental conditions of high pressures and temperatures. High Temperature — High Pressure **9** (1977) 6, 705–706.

[27] LEBEDEV, T. S.: Some problems of experimental petrophysical studies in the model thermo-

baric — conditions of the lithosphere. Geologische Korrelation — theoretische Grundlagen und praktische Erkenntnisfindung. Karl-Marx-Stadt (1979) 13—14.
[28] LEBEDEV, T. S.: Model studies of physical properties of mineral matter in high-pressure-temperatures experiments. Physics of the Earth and Planetary Interiors (1980) 25, 292 to 303.
[29] LEBEDEV, T. S.: Elastizität und elektrische Eigenschaften von Gesteinen bei hohen Temperaturen und Drücken. Gerlands Beiträge zur Geophysik **90** (1981) 5, 373—384.
[30] LEBEDEV, T. S.; SHAPOVAL, V. I.; KORCHIN, V. A.: Effect of high temperatures and pressures on changes in ultrasonic velocities in some igneous and metamorphic rocks of the Ukrainian Shield. Gerlands Beiträge zur Geophysik **79** (1970) 2, 108—116.

Effective elastic moduli of isotropic rocks at high pressure

I. O. Bayuk, V. A. Kalinin

The experimental investigation of rocks at high pressure gives the most reliable information about their elastic properties. However, experimental methods are labour-consuming but there is a great number of rocks. Moreover, when the pressure rises, the size of sample decreases, so the specimen of rock might be unrepresentative for this rock. Therefore, theoretical determination of rock's elastic properties at high pressure is necessary.

It is known that the rock is a polycrystalline aggregate consisting of crystallites, which belong to different mineral phases and so have different shape and size. Therefore, elastic properties of rock are determinated in general by elastic properties of minerals forming this rock and also by space orientation and mutual disposition of its grains. It is macroscopic properties, which is of great interest for geophysical research. Therefore, the aim of these investigations is the determination of the effective elastic tensor C^*, which ties together macroscopic fields of strain ε and stress σ [1]:

$$C^* = \bar{\sigma}^{-1}\bar{\varepsilon}, \tag{1}$$

where $\bar{\sigma} = \frac{1}{\Delta V}\int\limits_{\Delta V} \sigma(\bar{r})\,\mathrm{d}\bar{r}$ and $\bar{\varepsilon} = \frac{1}{\Delta V}\int\limits_{\Delta V} \varepsilon(\bar{r})\,\mathrm{d}\bar{r}$ are space averages for $\Delta V \gg l^3$, where l the average size of mineral grain.

As rock is a microheterogeneous medium, in which space orientation, crystallografic crystallite's axis and also its mutual disposition in the rock are of random character, that it is natural to apply for demination of tensor C^* methods, which are based on a theoretical probability approach. In terms of this approach the Hook's law is:

$$\begin{aligned} \langle\sigma\rangle &= \langle C\rangle\,\langle\varepsilon\rangle + \langle C'\varepsilon'\rangle, \\ \langle\varepsilon\rangle &= \langle S\rangle\,\langle\sigma\rangle + \langle S'\sigma'\rangle. \end{aligned} \tag{2}$$

Here $\langle x\rangle = \int\limits_{-\infty}^{\infty} x f(x)\,\mathrm{d}x$ is the statistical average, where x is one of the values σ, ε, C, S ($S = C^{-1}$), $f(x)$ is the probability density of x, $x' = x - \langle x\rangle$ is fluctuation. The members $\langle C'\varepsilon'\rangle$, $\langle S'\sigma'\rangle$ accounted for the heterogeneity of material as they contain the fluctuations C', S' and also the peculiarities in disposition of grains, as they contain the values ε' and σ'.

When the results of statistical averaging coincide with the results of averaging by volume there is ergodicity [2]. Ergodicity takes place when the fields of strain and stress are macroscopical homogeneous, and fields C and S are statistical homogeneous (that is statistical characteristics of fields C and S depend only on radius-vector $\bar{r}$ joining two points in space but do not depend on the coordinates of these points). Obviously, the ergodicity takes place for rocks when tensor C may be determinated by (2). At present in geophysics the method for calculating the elastic moduli of rocks is widely used, based on the averaging by Voight-Reuss-Hill, which finally leads to neglecting the members $\langle C'\varepsilon'\rangle$ and $\langle S'\sigma'\rangle$. According to this method, the calculations are carried out in two steps [3].

In the first step it is assumed that the strain and stress in crystallite of one and the same mineral phase is equal (approaches of Voight and Reuss accordingly) and so we receive the average values of tensor's components

$$\begin{aligned}\langle C^{\nu}_{ijkl}\rangle &= \frac{1}{8\pi^2}\int_0^{\pi}\int_0^{2\pi}\int_0^{2\pi} C^{\nu}_{ijkl}(\theta,\varphi,\psi)\sin\theta\,\mathrm{d}\theta\,\mathrm{d}\varphi\,\mathrm{d}\psi,\\ \langle S^{\nu}_{ijkl}\rangle &= \frac{1}{8\pi^2}\int_0^{\pi}\int_0^{2\pi}\int_0^{2\pi} S^{\nu}_{ijkl}(\theta,\varphi,\psi)\sin\theta\,\mathrm{d}\theta\,\mathrm{d}\varphi\,\mathrm{d}\psi.\end{aligned} \tag{3}$$

Where θ, φ, ψ — Eulerian angles, ν — number of mineral phase. Then the elastic moduli is calculated for each mineral phase [3]

$$K_V{}^{\nu} = f_1(\langle C^{\nu}_{ijkl}\rangle),\quad \mu_V{}^{\nu} = f_2(\langle C^{\nu}_{ijkl}\rangle),\quad K_R{}^{\nu} = f_3(\langle S^{\nu}_{ijkl}\rangle),\quad \mu_R{}^{\nu} = f_4(S^{\nu}_{ijkl}). \tag{4}$$

According to Hill's unequalities [4]

$$K_R{}^{\nu} \leqq K^{\nu} \leqq K_V{}^{\nu},\qquad \mu_R{}^{\nu} \leqq \mu^{\nu} \leqq \mu_V{}^{\nu} \tag{5}$$

one assumes

$$\begin{aligned}K^{\nu}_{\mathrm{VRH}} &= (K_V{}^{\nu} + K_R{}^{\nu})/2,\\ \mu^{\nu}_{\mathrm{VRH}} &= (\mu_V{}^{\nu} + \mu_R{}^{\nu})/2.\end{aligned} \tag{6}$$

Elastic moduli calculated by (6) are called "average moduli by Voight-Reuss-Hill" (VRH). In the second step averagings by the rule of mechanical mixture are carried out:

$$\begin{aligned}K_1{}^{\mathrm{av.}} &= \sum_{\nu=1}^{N} V_\nu K^{\nu}_{\mathrm{VRH}},\quad \mu_1{}^{\mathrm{av.}} = \sum_{\nu=1}^{N} V_\nu \mu^{\nu}_{\mathrm{VRH}},\\ K_2{}^{\mathrm{av.}} &= \left(\sum_{\nu=1}^{N}\frac{V_\nu}{K^{\nu}_{\mathrm{VRH}}}\right)^{-1},\quad \mu_2{}^{\mathrm{av.}} = \left(\sum_{\nu=1}^{N}\frac{V_\nu}{\mu^{\nu}_{\mathrm{VRH}}}\right)^{-1}\end{aligned} \tag{7}$$

where V_ν is the volume concentration of mineral phase ν. On the basis of unequalities

$$K_2{}^{\mathrm{av.}} \leqq K \leqq K_1{}^{\mathrm{av.}},\qquad \mu_2{}^{\mathrm{av.}} \leqq \mu \leqq \mu_1{}^{\mathrm{av.}} \tag{8}$$

it is put as

$$K = (K_1{}^{\mathrm{av.}} + K_2{}^{\mathrm{av.}})/2,\qquad \mu = (\mu_1{}^{\mathrm{av.}} + \mu_2{}^{\mathrm{av.}})/2. \tag{9}$$

One can see that the members $\langle C'\varepsilon'\rangle$, $\langle S'\sigma'\rangle$ may be presented by the sum of two components, which characterize the anisotropy of minerals belonging to various mineral phases. It is known that the most of rock-forming minerals are anisotropic [5] and the values of their elastic moduli may differ from each other by several times [6].

So neglecting the members $\langle C'\varepsilon'\rangle$ and $\langle S'\sigma'\rangle$ may lead to great mistakes in the effective moduli of rock. Among the theoretical probability methods, which allow to consider these members approximately, the method of generalized singular approach (GSA) exists [7]. This method is the most acceptable for rocks because it

1. does not put any restriction on the crystalline class of symmetry;
2. does not demand the proximity of elastic properties of different mineral phases;
3. does not put any assumption about the shape of grains (for example, spherical or ellipsoidal shape).

This method is based on the comparison of strain's and stress's fields, which arise in heterogeneous body under stress and in homogeneous body with known elastic properties ("comparison body"). The effect of mutual disposition of grains on the strain's and stress's fields is accounted by the Green's function of "comparison body". However, the members in this method may be accounted only approximately, thus the values of strain in grains belonging to one and the same mineral phase with the identical orientation of crystalline axis are equal.

Let us consider macroisotropic rock. In this case the formula for calculations by the method GSA is [7]

$$(C^* + b^c)^{-1} = \langle (C^c + b^c)^{-1}\rangle \tag{10}$$

where $b^c = 3b_k{}^c V + 2b_\mu{}^c D$, $b_k{}^c = \frac{4}{3}\mu_c$, $b_\mu{}^c = \frac{\mu_c(9K_c + 8\mu_c)}{6(K_c + 2\mu_c)}$, V and D are volume and deviator components of unit tensor, respectively, K_c, μ_c are elastic bulk and shear moduli of "comparison body". The index "c" denotes values characterized "comparison body".

In this work the formula (10) has been extended to the case of multiphased polycrystals, which the rocks are in reality. It has been proved, that the effective moduli of isotropic rocks may be calculated by the next formulas

$$\begin{aligned} K^* &= \left\{\sum_{\nu=1}^{N} V_\nu (K^\nu + b_k{}^c)^{-1}\right\}^{-1} - b_k{}^c, \\ \mu^* &= \left\{\sum_{\nu=1}^{N} V_\nu (\mu^\nu + b_\mu{}^c)^{-1}\right\}^{-1} - b_\mu{}^c. \end{aligned} \tag{11}$$

Based on these expressions we can conclude what is of great practical importance: for the calculation of elastic moduli of isotropic rock one can use the experimental data on the polycrystalline specimens of minerals instead of the experimental data on complete elastic tensors of these mineral phases.

Furthermore, from the structure of the formula (11) we can say that the effective moduli of rocks which contain liquid ($\mu = 0$) or gaseous ($K = 0$, $\mu = 0$) inclusions are calculable. It is impossible to do so if we use the method based on the averaging by VRH (below we shall say "averaging by VRH"). So, if we know from experiment the porosity of rock and its mineral composition, we can calculate by applying the GSA method its elastic moduli considering the effect of spherical porosity.

The results of calculations according to the GSA method have been compared with

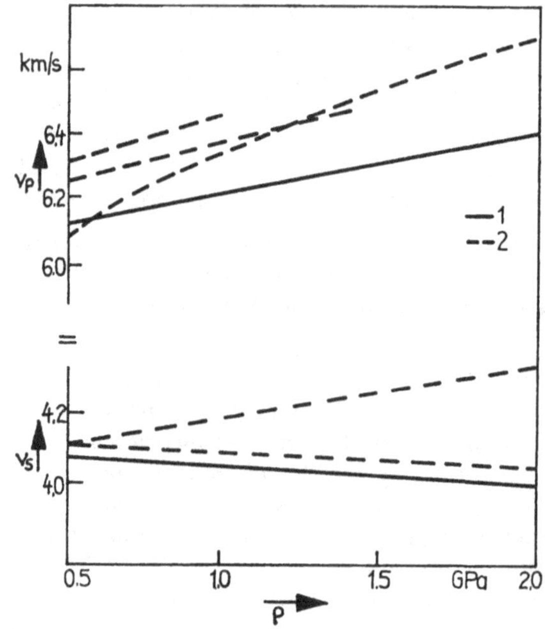

Fig. 1. Dependence of the compressional and shear velocities on pressure for the quartzite. *1* – the results of the calculations by the method GSA, *2* – the experimental data of different authors [9–11].

the experimental data. In Fig. 1 there are the velocities of compressional and shear waves in quartzite. The thick line represents the velocities calculated by GSA, the other lines are experimental data of different authors [9–11]. The maximum difference between calculated and experimental values is less than 2.5%.

The analogical comparison has been made for isotropic granite Westerly, which consists of quartz – 28%, oligoclase – 36.6%, microcline – 36.3%, and biotite – 3.3% [12–15]. Experimental data for these minerals have been taken from [11, 16, 17]. We have investigated the effect of spherical porosity on the elastic moduli of this rock. In Fig. 2 the solid line represents the results of calculation for the porous Westerly granite, which has 1.2% of spherical pores [12]. In this figure we can see a good agreement of velocities calculated by GSA with the experimental values. Without accounting the porosity the maximum difference in velocities is 3%, but with the account of pores – 1.7%. This difference is decreasing with increasing pressure. The maximum difference between the calculations by VRH and the experimental data is more than 3%, that exceeds the accuracy of measurements – 1–2%.

So the GSA method can be used more widely than VRH so it is applicable for the rocks containing minerals which have considerable anisotropic elastic properties, and the values of the elastic moduli of mineral phases may be of great difference from each other.

Furthermore in contrast to VRH, the GSA method allows to account the effect of spherical porosity.

Moreover the extent of calculations by GSA does not exceed the corresponding extent by VRH.

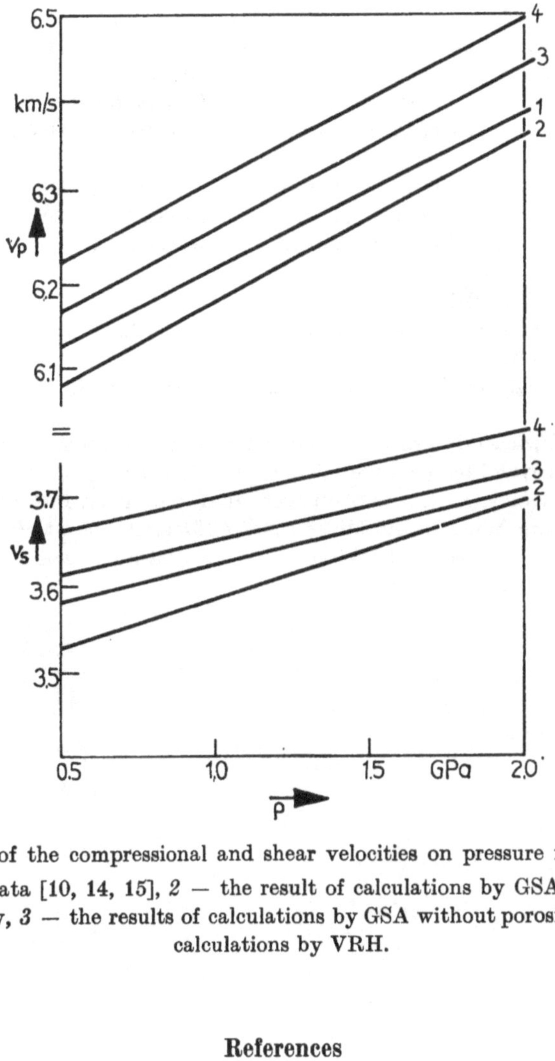

Fig. 2. Dependence of the compressional and shear velocities on pressure for Westerly granite. *1* — experimental data [10, 14, 15], *2* — the result of calculations by GSA with the account of the effect of porosity, *3* — the results of calculations by GSA without porosity, *4* — the result of calculations by VRH.

References

[1] Horoshun, L. P.: Mechanika kompozitnih materialiv i elementov konstrukci. „Naukova dumka" 1 (1982) 176—190.

[2] Lomakin, V. A.: Statisticheskie zadachi mehaniki tverdih deformiruemih tel. „Nauka" (1970) 138 p.

[3] Belikov, B. P.; Alexandrov, K. S.; Rhyzhova, T. V.: Uprugie svoistva porodoobrazujshih mineralov i gornih porod. „Nauka" (1970) 276 p.

[4] Hill, R.: The Elastic Behaviour of a Crystalline Aggregate. — Proc. Phys. Soc., A **65** (1952) 389, 349—354.

[5] Volarovich, M. P.; Bayuk, E. I.; Efimova, G. A.: Uprugie svoistva mineralov pri visokih davlenijah. „Nauka" (1975) 131 p.

[6] Anderson, O. L.; Schreiber, Ed.; Liebermann, R.; Soga, N.: Some Elastic Data on Minerals Relevant to Geophysics. Rev. Geoph., **6** (1968) 4, 491—523.

[7] Shermergor, T. D.: Teorija uprugosti microneodnorodnih sred. „Nauka" (1977) 400 p.

[8] Volarovich, M. P.; Levikin, A. I.: Vlijanie davlenija do 40 kbar na uprugie svoistva nekotorih porodoobrazuyuschih mineralov. Geoph.sb., **54** (1973) 35—38.
[9] Chung, D. H.; Simmons, G.: Pressure Derivatives of the Elastic Properties of Polycrystalline Quartz and Rutile. — Earth and Planet. Sci. Letters, **6** (1969) 2, 134—138.
[10] Birch, F. G.: The Velocity of Compressional Waves in Rocks to 10 kbar. J. Geoph. Res., **65** (1960) 4, 1083—1102.
[11] McSkimin, H. Y.; Andreatch, P.; Thurston, P. N.: Elastic Moduli of Quartz Versus Hydrostatic Pressure at 25° and 195,8 °C. J. Appl. Phys., **36** (1965) 5, 1624—1632.
[12] Walsh, J. B.: The Effect of Cracks on the Compressibility of Rocks. J. Geoph. Res., **70** (1965) 2, 381—389.
[13] Ultrazvuk v geofizike. „Mir" (1964) 274 p.
[14] Birch, F.: The Velocity of Compressional Waves in Rocks to 10 kbar. J. Geoph. Res., **66** (1961) 7, 2199—2224.
[15] Simmons, G.: Velocity of Shear Waves in Rocks to 10 kbar. J. Geoph. Res. **69** (1964) 6, 1123—1130.
[16] Galdin, N. E.: Fizicheskie svoistva glubinnih metamorficheskih i magmaticheskih porod pri visokih davlineyah i temperature. „Nedra" (1977) 127 p.
[17] Lebedev, T. S., Korchin, V. A., Orovetsky, Ju. P.; Burtny, P. A.: Elastic Properties of Some Rock-forming Minerals and Mineral Matter Deformation at High Hydrostatic Pressure. In: Physical Properties of Rocks and Minerals Under Extreme *P,T*-conditions. Akademie-Verlag, Berlin (1979) 31—41.

Structural transformations of rocks and their effect on changes in elastic properties under different thermobaric conditions

T. S. Lebedev, G. Y. Novik, V. A. Korchin, M. G. Silbershmit, T. K. Zavorikina

The inhomogeneities of lithosphere's mineral matter elastic property variations with depth are caused by many geological and geophysical factors. Their role is only partly known for the most cases even now. The ultradeep wells doubt the existence of some seismic boundaries whose nature has been associated with compositional changes of media alone. Thermobaric studies of physical properties of rocks by experiments imitating P-T regimes in different parts of the lithosphere together with wave- and other geóphysical as well as geological data enable as to predict possible changes of the mineral matter state and composition and to specify the interior structure. Some progress has been made in this field but the data obtained are scare as yet.

The simultaneously programmed laboratory high P-T studies of mineral matter under conditions similar to real deep ones reveal new features of distribution of its elastic parameters. They are manifested in zones of extreme rock-type-dependent longitudinal (V_p) and transverse (V_s) elastic wave values as well as P-T ratios and gradients [1—5].

As an example most specifialy modelled $V_p, V_s = f(P, T) = f(H)$ dependences of some typical granitoids of the Ukrainian Shield (USh) are briefly considered [1—5] without detailed discussion of experimental data sets.

The programmed P-T experiments as well as some velocity isobar and isotherm calculations show $V_p, V_s = f(P, T) = f(H)$ dependences to differ significantly from those observed when rocks are effected by pressure alone at room temperature (Fig. 1). $V_p = f(P)$ curves show linear velocity increase at $P = 0.1$ GPa, whereas $V_p, V_s = f(P, T) = f(H)$ ones illustrate complex velocity change laws. In the P-T range corresponding to depths of the upper 2—4 km the velocity values increase as they do under the effect of hydrostatic pressure at 20 °C. Further P-T increase, i.e. "plunging" of the samples to 5—10 km first lowers the velocity which then increases again in many cases showing maxima at depths of 15—25 km. Thus, several extreme zones are distinguished on the velocity curves (Fig. 1). greatly depending on the character of P and T distribution at depths as well as additional tensions. Note that pressure decrease from its mean calculated value for respective depth at $T > 150$ °C expands velocity inversion zones and decreases their V_p and V_s more intensively. Pressure increase acts inversely as do temperature deviations from the mean value at respective pressure in experiments.

Let us consider the informativity of elasticity module variations for mineral matter calculated from experimental V_p, V_s and ϱ values for different depths lets as a definite

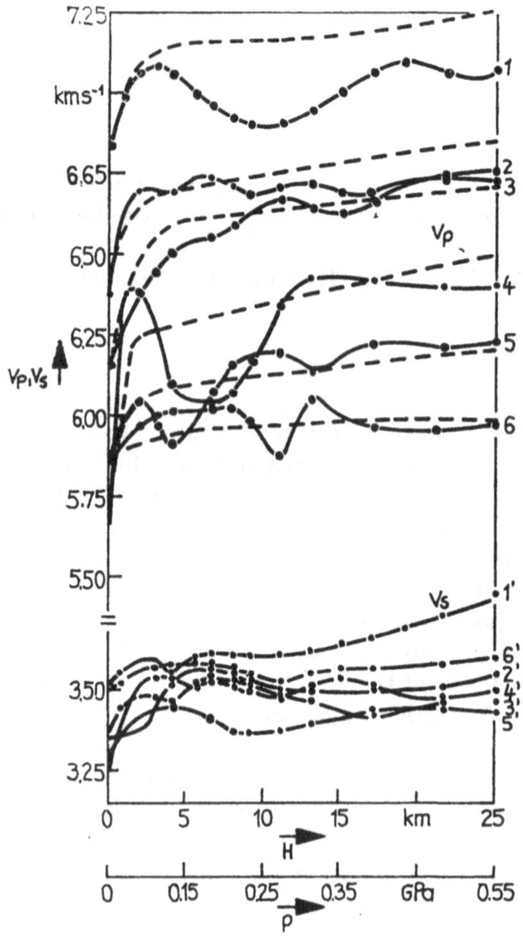

Fig. 1. V_p and V_s variations in Ukrainian Shield granites.

1 — pegmatoid, *2* — porphyry (new-ukrainian), *3, 5* — rapakivi, *4, 6* — medium-grained (lelekovo area).

Solid lines: V_p, $v_s = f(P_q, T)$, dashed lines: $V_p = f(P_h)$.

macrosystem whose general feature is higher elasticity of mineral matter (high Young's (E) and shear (G) moduli) in high-velocity zones (compared to low-velocity ones) with its low compressibility (β) and sufficient plasticity (high Poisson's ratios (μ)). Low-velocity zones generally contain brittler (low μ), more friable (high β) and low-elasticity (decreasing E and G) rocks. Such involved response of the mineral macrosystems to the programmed effect of P and T is doubtless not only due to their effect on thermo-elastic state of minerals but also to definite structural transformations of the whole rock or its parts.

To expand our knowledge as to where extreme zones of elastic wave velocity variations originate from at different depths modelled several experiments aimed at finding their relation to possible structural transformations and associated residual stress resulting from active thermobaric effects have been made.

As known, P and T increase changes the initial structure of the entire mineral substance which is associated with the effect of thermodynamic parameters on the structure of anisotropic rock-forming minerals. The change of the physical state of rocks under any external P-T effects manifest themselves at micro- and macro-levels, the kinetics and sequence of these changes being intimately related to substance structure and composition as well as to external fields of different types and regimes acting on them.

To elucidate the nature of these processes and phenomena all aspects of the rock as an anisotropic medium should be analysed considering its polycrystalline multicomponent features and the general character of its mineral components.

Such a method enables us to study the changes in the crystal lattice of minerals, inter-grain space and the whole macrovolume of rocks combining theoretical and experimental data. In particular, features of electric stress field of a rock exposed to various external field effects may be estimated from the view-point of mechanics of a micro-inhomogeneous medium with prognosing possible structural changes of its mineral components studied by such structure-sensitive methods as X-ray diffractometry, optical and electronic microscopy. Quantitative structural analyses of the samples used need data sets indicating differences of their destruction at different level and tension of their elements.

We study one of the USh granitoid varieties whose elastic features had been thoroughly studied under different thermobaric conditions. 10 cylindric samples of grey equigranular granite were investigated each about 22 mm in length and diameter and bored in one direction.

The mean grain size of the rock-forming mineral samples varies from 0.3 to 0.8 mm, their content being $32 \div 35$, $20 \div 27$, $29 \div 39$, $6.8 \div 9$, $1.7 \div 3.0$ volume-% of quartz, potassium feldspars, plagioclases, biotite and accessories respectively.

The mean volume weight of the rock and its mineral density are 2.664 and $2.683 \cdot 10^3\ kg \cdot m^{-3}$. The porosity varies from 0.7 to 1.1%. The mean V_p and V_s values at atmospheric pressure are $4920\ m \cdot s^{-1}$ (min — 4870, max — 4950) and $3040\ m \cdot s^{-1}$ (min — 3000, max — 3100), respectively. At axial pressure of about 4 MPa they are $5160\ m \cdot s^{-1}$ (min — 5110, max — 5190) and $3190\ m \cdot s^{-1}$ (min 3110, max — 3230), respectively. For a single sample, longitudinal and transversal wave anisotropy did not exceed 2%. V_p, V_s and ϱ values as well as thin section data evidence high homogeneity of minerals studied, the differences of elastic features correlating well with original structural features of rock-forming minerals.

Thermobaric changes of elastic parameters

For special experiments four granite samples were chosen. Each of them was studied by a model P-T programme (Fig. 2). Fig. 3 and 4 show respective results of thermobaric V_p and V_s variations as well as calculated Young and shear modules, compressibility and Poisson's ratio. Sample 1 was studied at room temperature and hydrostatic pressure to 0.52 GPa, the pressure transmitting medium is spindle oil (Fig. 3, 4, sample 1), whereas sample 2 was exposed to hydrostatic pressure to 0.52 GPa (using lead as

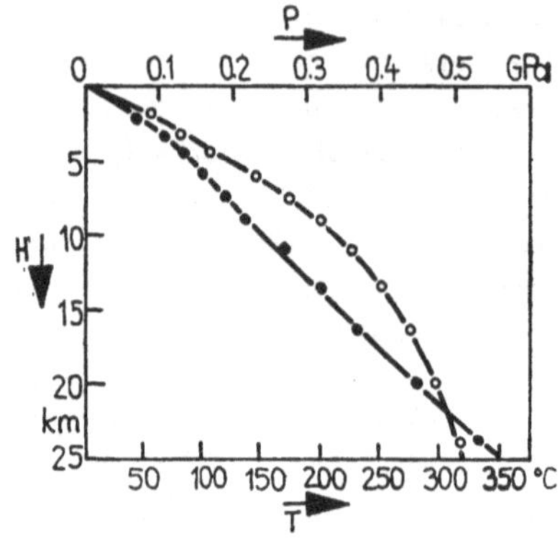

Fig. 2. Programme of model *P-T* experiments.

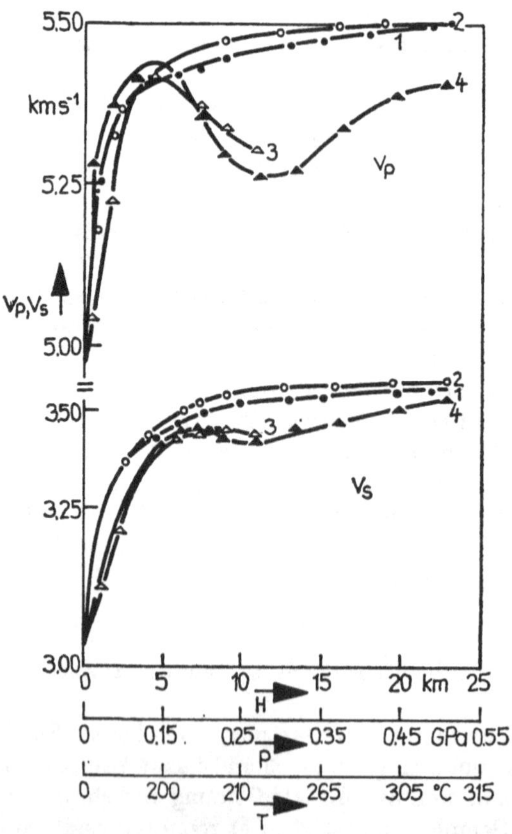

Fig. 3. V_p and V_s changes in equigranular granite at hydrostatic (sample *1*) and quasi-hydrostatic (sample *2*) pressure as well as in programmed *P-T* modelling regime (samples *3*, *4*).

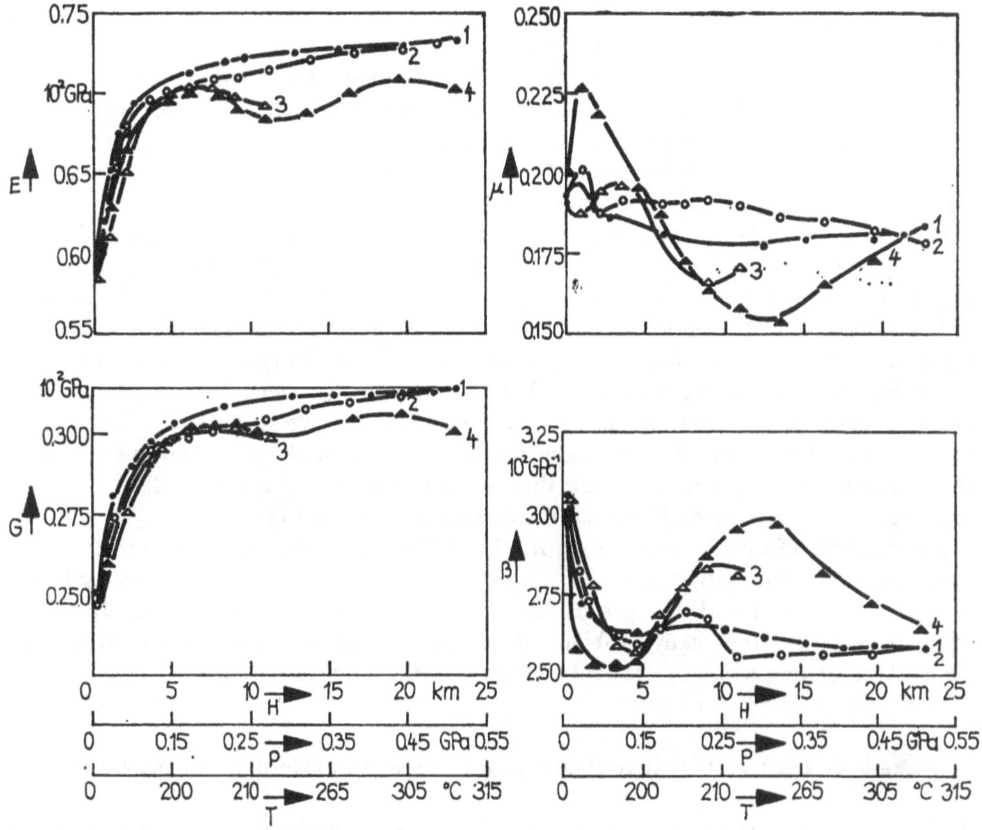

Fig. 4. Changes in Young (E) and shear (G) module, Poisson's ratio (μ) and compressibility (β) of equigraniular granite at hydrostatic (sample *1*) and quasi-hydrostatic (sample *2*) pressure and by model *P-T* program (samples *3*, *4*).

pressure transmission medium) and room temperature (Fig. 3, 4). Sample 3 (Fig. 3, 4) was simultaneously effected by quasi-hydrostatic pressure (to 0.27 GPa) and temperature (to 225 °C) according to model *P-T* program (Fig. 2). Finally, sample 4 was treated in essentially the same programmed thermobaric regime to 0.52 GPa and 310 °C (Fig. 3, 4). The same sample effected by *P* and *T* was used to measure longitudinal and transversel wave velocity and volume decrement. The experimental methods and apparatus have been prepared by Earth's substance physical properties Department Inst. of Geophys. Ac. Sci. Ukr. SSR (Kiev) [2, 6—8]. The relative V_p and V_s variation determination error did not exceed 1%.

The experimental data analysis shows the changes of elastic features of the samples studied to depend primarily on pressure category (hydrostatic or quasi-hydrostatic) and temperatures values at corresponding pressures. Quasi-hydrostatic pressure compared to the hydrostatic one) is just that yielding the most intensive elastic wave velocity increase with pressure elevation at room temperature. Here velocity hysteresis is also observed with pressure application and removing. Typical for quasi-hydrostatics are rather intensive compressibility changes and higher sample plasticity expressed qualitatively in corresponding μ behaviour though the absolute Poisson's ratio value of the rock studied suggests it to be brittle.

Thus, the comparison of changes in V_p, V_s and other elastic features under hydrostatic and quasi-hydrostatic conditions shows that non-hydrostatic pressure yields additional tension throughout the samples. It activates structural changes and compresses mineral matter more intensively. This conclusion is detailized below using concrete data on structures of the samples studied.

Programmed model experiments on sample 4 (Fig. 3) revealed complicate thermobaric V_p and V_s relations expressed in the clearcut V_p and slightly less pronounced V_s peak at P and T corresponding with depths of circ. 4 km and a distinct low-velocity zone in the 0.16—0.45 GPa and 175—290 °C range, i.e. at depths to 20 km (Fig. 3). Similar results are typical of samples 3, effected by P and T according to an incomplete program (Fig. 2, 3). Analysing the laws of E, G, μ and β changes we note that in the initial pressure and temperature range the elastic parameters behave so as they do in the case of the effect of pressure alone at room temperature. Then the simultaneous P and T effect slightly discompacts the rock diminishing Young and shear modules with increasing compressibility and notable Poisson's ratio decrease (Fig. 4).

The analysis of elastic parameters in this PT-range shows that pressure does not compensate the destructive action of temperature and is expressed in weakening bonds at grain boundaries and changes in rock-forming mineral microstructures. It is just what we consider the main problem of studying residual phenomena in structural elements of granite samples exposed to high P and P and T.

Methods and results of investigations of structural transformations in granite

X-ray and optical studies were made on polished sections of granite samples for which PT-variations of respective physical characteristics were studied. The same sample surfaces were analyzed before and after the pressure and temperature effect. Block size within rock-forming minerals were studied, residual deformation and changes of dislocation density in them, residual stress concentrations of deformation defects and twins as well as dislocation density in inter-block spaces.

In addition to the above structural data, elastic stress field inhomogeneities were calculated for granite samples in different PT-regimes. For single elements of the rock considered to be a quasi-isotropic multicomponent polycrystalline aggregate consisting of potassium feldspar (microcline), plagioclases and quartz in proportion to corresponding mean real statistic values, local stress was estimated with respect to mean macrovalues for concrete PT-conditions. The calculations use equations of statistic mechanics of microinhomogeneous media [9] and reference values of physical tensors of spherical characteristics of respective rock-forming minerals [10]. All the rock grains are shown to be compressed at hydrostatic pressure. However, the mean stress values in plagioclase grains exceed the mean microstress values of the whole rock by 15—20%, this stress being by 10—15% lower for quartz grains. The thermal effect yields somewhat different stress nonuniformities. In particular, the highest compressional stress should theoretically be observed in quartz grains $\perp$ to $\bar{c}$-axis, while in microcline ones they should occur $\parallel$ to $\bar{a}$. The highest tensile stress is expected in microcline ($\parallel$ to $\bar{b}$ and $\bar{c}$) and plagioclase ($\parallel$ to $\bar{b}$) grains. At simultaneous PT-effect, the stress state inhomogeneities are most probably due to superposition of fields of elastic deformations caused by these two factors.

Theoretical calculations show that in concrete *PT*-conditions the local threshold stress of non-elastic microdeformations in potassium feldspars and plagioclases is lower than in quartz. Hence, the structural changes of the latter may be used as a criterion of estimation of the general structural state of a rock and a means enabling us to avoid great difficulties in studying other rock-forming minerals with less symmetric crystal structure. Therefore, X-ray studies were just made on quartz grains.

To determine mineral fine-structure parameters by X-ray diffractometry approximate intensity distribution method was used since comparative analysis of mineral structure parameter variations is not less accurately made by Fourier-transformated approximating functions than by harmonic analysis [11].

The calculation of rock-forming mineral structure parameters was made in three stages.

1. Preparing diffractograms of standard samples and those under study. Determination of the area below the curves of the same interference line of the standard sample and that under study. Dividing respective areas by curve peak height yields the integral half-width B (for the sample under study) and b (for the standard sample). The broadening is expressed in radian measure with taking diffractogram size into account. Picturing by characteristic radiation needs corrections for doubling spectral lines K, α_1 and α_2 since the interference line obtained is formed by superimposing independent α_1- and α_2-curves. The separation is made by Reshinger method [12]. Then correction for instrumental broadening is introduced. Thus

$$\beta_b = B - \frac{b}{B}, \tag{1}$$

where β_b is physical broadening.

2. The separation of the effects associated with small dimensions of the coherent scattering areas and degree of microdeformation. Note that intensity distribution across the line dissociated by crystal lattice distortions is approximated by Gauss function and by Cauchi one due to lattice block structure.

For β_b values of various diffractogram lines straight lines are obtained by calculations based on $\beta_b \cos(\theta) = f(\sin\theta)$ dependence (θ — angular position of difracted line). The intersection of these straight lines with the coordinate axis yields λ/D values (where λ and D are wavelength and block dimension respectively). To determine the mean residual deformation values for blocks $\varepsilon = \frac{\Delta a}{a}$ (where a is quartz parameter lattice) $\beta_b^2 \cos^2\theta = f(\sin^2\theta)$ curves are set up. If the slope angle tangent with respect to abscissa is known we find ε [13, 14]. The sign of relative deformations $\frac{\Delta a}{a}$ or residual stresses σ characterizes the state of single unchanged quartz grains in stress fields. A compressed crystal is given 2 "—" while that extended beyound the uncharged limits suggests residual stress and deformation with "+". This follows from calculated proportions for σ and $\frac{\Delta a}{a}$ omitted here due to their cumbersome character.

3. Determination of deformational (α) and twin (γ) packing defect concentration as well as dislocation density depending on mosaic blocks for hexagonal dense packing structure. The most accurate results of the calculations of concentration of deformational defects are obtained as follow

$$\alpha = \frac{\left(\frac{\sin\theta_i}{\sin\theta_j}\right)_{\text{def}} - \left(\frac{\sin\theta_i}{\sin\theta_j}\right)_{\text{nondef}}}{\frac{1{,}36}{\pi^2}\left[\left(\frac{\sin\theta_i}{\sin\theta_j}\right)_{\text{nondef}} \times A(H_1K_1L_1) - \left(\frac{\sin\theta_i}{\sin\theta_j}\right)_{\text{def}} \times A(H_2K_2L_2)\right]}, \tag{2}$$

where $A(HKL) = \sum^{w} (\pm L_0)/(u + w)\, h_0^2$; $L_0 = H + K + L$; $h_0^2 = H^2 + L^2 + K^2$; H, K, L are indices of lines in cubic axes; u is number of components of lines which are not deviated by packing defects; w is the number of the components of lnes which are deviated; θ_i, θ_j are Braggian angles of line peaks i and j.

The concentration of twin packing defects may be calculated from the equation

$$(1{,}5\alpha + \gamma) = \frac{a}{C(H_1K_1L_1) - C(H_2K_2L_2)} \times \frac{K - 1}{D(H_1K_1L_1)}, \tag{3}$$

where $C(HKL) = \sum^{w} |L|_0/(u + w)\, h_0$; $k = D(H_2K_2L_2)/D(H_1K_1L_1)$; D — block size; $D(HKL) = 0.84\lambda/2\beta_b \cos\theta$.

The dislocation density (ϱ_D) in the interblock space is detected by formula $\varrho_D = \frac{3n}{D^2}$, where n is number of dislocations on each block face.

When dislocations are distributed chaotically the dislocation number n is taken to be 1.

The dislocation density in quartz grain blocks is determined by $\varrho_\sigma = \frac{q}{b^2}\,\bar{\varepsilon}^2$ where q is the proportionality coefficient depending on elasticity and shear modules and b is the Burge s vector.

The structural changes of the rock effected by P and T were analysed on the basis of the quartz grain state. The results of quantitative estimation of relative changes in structural parameters of quartzs are given in Table 1 and 2. X-ray diffractometry provided data on block size in grains (D), their relative deformation ($\Delta a/a$), dislocation density in blocks (ϱ_δ), residual stress (σ), concentration of packing defects and twins ($1.5\alpha + \gamma$) and the dislocation density in inter-block space (ϱ_D).

In optical studies special methods of sample surface preparation by chemical etching were used. The surfaces were studied by microequipment MKE-I in the light and dark field.

The etching revealed single fragments of potassium feldspar and plagioclase grains and defined clearly their boundaries. For some parts of the surfaces studied select sighting surveys providing visual data on changes in samples effected by P-T were made. E.g. quartz-grain boundaries were found to be clear-cut and showed no visible changes; somewhat transformed were, however, quartz—feldspar and quartz—biotite boundaries. Feldspars were disintegrated and crushed along the boundaries. Potassium—feldspar grains (seen in dark field) split into blocks. They could also be seen structural changes of some feldspar grains of which the main is disappearance of twin bands. No gliding lines were seen in biotite. Broadening of biotite—feldspar boundary was observed.

Table 1. Structural parameters and residual stress of quartz grains in granite samples effected by P and T

Sample number	P-T regimes	Block size, D (Å)	Residual deformation, $\Delta a/a$ (10^3)	Dislocation density in blocks, ϱ_σ (10^7, Å^{-2})	Residual stress, σ (10^3 GPa)	Concentration of deformational defects and twins	Density of interblock dislocations, ϱ_D (10^7, Å^{-2})
1.	Initial	515.5	+3.994	12.05	+11.56	3.66	89.58
	0.52 GPa — hydrostatics	436.8	+4.434	15.14	+12.96	3.56	124.80
2.	Initial	2178.1	−1.4104	1.533	−4.124	1.061	3.82
	0.52 GPa — quasihydrostatics	1975.7	−2.215	3.765	−6.464	2.055	1.15
3.	Initial	730.4	+2.393	4.337	+6.938	1.613	44.60
	0.27 GPa — quasihydrostatics, 225 °C	657.6	+2.462	4.681	+7.207	1.469	55.02
4.	Initial	746.3	+2.682	5.545	+7.845	2.297	42.75
	0.52 GPa — quasihydrostatics, 310 °C	975.0	−0.602	0.279	−1.759	1.068	24.99

Table 2. Relative changes of structural parameters and residual stress of quartz grains in granite samples effected by P and T

Sample number	P-T regimes	Block size, D (Å)	Residual deformation, $\Delta a/a$ (10^3)	Dislocation density in blocks, ϱ_b (10^7 Å^{-2})	Residual stress, σ (10^3 GPa)	Concentration of deformational defects and twins	Density of interblock dislocations, ϱ_D (10^7 Å^{-2})
1.	0.52 GPa — hydrostatics	0.85	+0.400	+1.25	+1.12	0.97	1.39
2.	0.52 GPa — quasihydrostatics	0.89	+0.805	+2.44	+1.56	1.92	0.40
3.	0.27 GPa — quasihydrostatics, 225 °C	0.90	+0.069	+1.08	+1.04	0.91	1.23
4.	0.52 GPa — quasihydrostatics, 310 °C	1.30	−3.284	−19.87	−1.22	0.47	0.58

The hydrostatic pressure effect (to 0.52—0.55 GPa) at room temperature caused non-uniform elastic stress in rocks associated with changes and anisotropy of elastic parameters of rock-forming minerals. As the maximum compressional stress is expected in plagioclase and potassium feldspar grains the latter are the most probable sources of non-elastic microdeformation. In particular, deformational twins are expected to manifest themselves in grain parts with the highest suitably oriented local stress. This process is very probable for feldspars at room temperature. Optical investigations show that the rock reconstructs twin structure of definite orientation (Fig. 5), the quartz grains changing slightly concentration of packing defects and twins together with residual stress increase (Table 1 and 2) and distorting their crystal structure by diminishing block size and increasing dislocation density in and between them.

Quasi-hydrostatic experiments (sample 2) exhibit additional stress in the chamber resulting in respective structural changes in the samples. They are much more intensive in all rock-forming minerals than those occurring in hydrostatics. Potassium feldspars and plagioclases twine alongside with quartz (Table 1 and 2). Shear load disturbs the inter-grain boundaries (Fig. 6). These processes increase stress in quartz and dislocation concentration in blocks.

X-ray and optical studies confirm the assumption of successive origin of elastic and plastic deformations of rock-forming minerals worsening the initial rock structure. Similar but more pronounced (compared to those seen in hydrostatics) changes occur in samples effected by quasi-hydrostatic pressure. The residual stress in granite samples effected by hydrostatic and quasi-hydrostatic pressure accords with changes in their elastic parameters, the first thing to be noted being that more intensive structural transformations at quasi-hydrostatic than at hydrostatic pressure compact the rock actively causing significant V_p and V_s changes. Note that $V_p, V_s = f(P_h)$ or $V_p, V_s = f(P_q)$ dependence (Fig. 3) and elastic parameter behaviour (Fig. 4) characterize dynamics of non-elastic processes in granite evidencing threshold character of its structural changes under confining pressure. These changes evidently occur and intensify at $P > 0.1$ GPa, i.e. within the range of sharp velocity gradient decrease.

A quite different picture of changes in physical state of granite is observed at simultaneous P and T effect with thermoactive processes taking place in rock-forming minerals (at microlevel) and the whole sample (as a macrosystem).

The entire rock structure is transformed by changes of stress of its structural elements due to anisotropy of their thermal expansion and compressibility as well as diffusive microdeformation.

Estimation based on theoretical calculations shows the highest compressional stress in feldspar grains to be relaxed by thermal effects $\parallel b$ and $\parallel c$ in microcline and $\parallel b$ in plagioclases, the most stressed quartz grains being in the direction $\perp c$. Moreover, thermal effect and growing additional stress in quartz create conditions activating its twinning by vacant-diffusive mechanism and partial clearing of defects and impurities. Thus, a threshold character of the structural changes of the rock may be also expected at PT-effect.

Comparing the sample 2 (quasi-hydrostatic pressure to 0.52 GPa at room temperature) and sample 4 (quasi-hydrostatic pressure to 0.52 GPa at 310 °C) data temperature contribution (at high pressure) to changes in mineral matter state at simultaneous PT-effect (Fig. 3 and 4; Table 1, 2) can be determined. In the first case (sample 2), the rock sharply deteriorates its structural features, which is clearly expressed in dis-

Fig. 5. Occurence of new twins in feldspar grains under hydrostatic pressure (sample *1*). 90 — fold magnification.

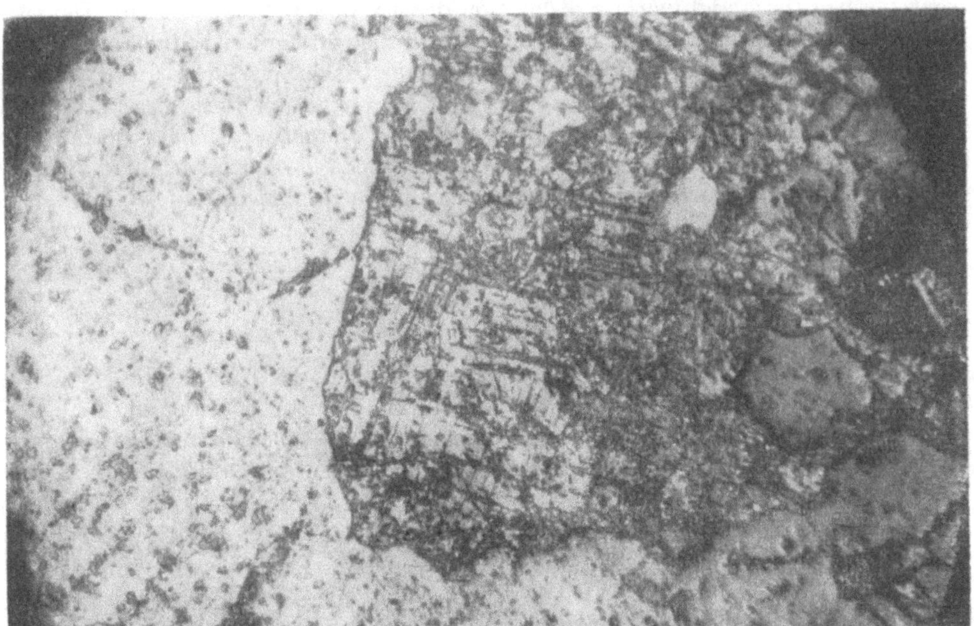

Fig. 6. Distortion of one of inter-grain boundaries in equigranular granite under quasi-hydrostatic pressure (sample *2*). 75 — fold magnification.

location density increase in quartz grain blocks, deformational defect and twin concentration, and increases its residual stress caused by growing non-elastic microdeformation in potassium feldspar and plagioclase grains. In the second case (sample 4), heating to 310 °C at the same pressure yields opposite structural changes.

This difference is especially clearly seen in quartz grains whose structure becomes much more perfect (Table 1). The block sizes increase by more than one order, the dislocation density decreases in blocks, this decrease becoming almost twofold between them. This is probably due to vacancy-diffusive processes occuring at non-elastic quartz deformations. However, a lot of structurally distorted feldspar grains (i.e. those with increased twin density) appear in the sample due to maximum thermal stresses. The expanding non-elastic processes in plagioclases and feldspars caused by temperature may explain the change of the sense of residual stress in quartz grains. Such complex P and T-transformations, especially favouring definite remanent structural perfection of rock evidently cause some changes in its elastic parameters in complicate thermobaric regimes. Though high temperature deteriorates elastic features of rock under normal conditions it exhibits high velocities at simultaneous PT-effect whose further increase with depth may be expected. The rock becomes more mobile, plastic, elastic and less compressible (Fig. 4), i.e. closer to microstructural perfection. Its transition to a qualitatively new state in respective PT-situations is most probably of threshold character. This is confirmed by sample 3 effected experimentally by P and T of 0.27 GPa and 225 °C respectively. Under such PT-conditions of V_p, $V_s = f(P, T) = f(H)$ dependences a low-velocity zone (Fig. 4, sample 3, 4) appeared in which, unlike the changes seen in sample 4, effected by $P = 0.52$ GPa and T to 310 °C the rock worsens slightly its initial structure (Table 1 and 2). The quartz grains diminish somewhat the block size and increase dislocation density in and between them. The twin density is also increased in feldspars broadening their intergrain boundaries. Similar changes were observed in granite samples exposed to high quasi-hydrostatic pressure at room temperature (sample 2), which means high ununiform stress growing in their PT-interval considered and which (together with thermoelastic ones) worsens its structural and elastic state significantly. Only further PT-increase acts like a trigger of thermoactive rock-perfection mechanisms at all structural levels.

Conclusions

Analysis of P and T effect on elastic features of granites changing their structural characteristics has revealed:

- correlation of initial elastic and micro-structural parameter distribution,
- significant deterioration of structural characteristics at high pressure and room temperature changing the gradient of elastic parameter increase,
- threshold character most probably carried by P and T effected structural changes,
- the most intensive of them occuring at P and T at depths of 12–15 km, i.e. defined before in the low-velocity zone,
- structural perfection of rock with further "plunging" of samples to thermobaric levels of 0.52 GPa and 310 °C resulting in velocity increase.

Thus, the laws of mineral matter elastic parameter distribution at different P and T are greatly influenced by changes in structured features of rock-forming minerals. They may be used as an indicator of processes occuring in the mineral matter (at respective natural and/or laboratory PT-combinations).

The above set of experiments made by new methods may be concluded to be a definite stage on the new way proposed. The future period should include even more detailed petrophysical studies together with special x-ray and optical researches of a wide range of minerals of different rocks which have been effected by P and T.

References

[1] Lebedev, T. S.: Problems of thermobaric studies of lithosphere's mineral matter physical properties (in Russian). Geofiz. Journ., 7 (1985) 6, 62–82.

[2] Lebedev, T. S.; Korchin, V. A.; Burtny, P. A.: Elastic wave velocities in ultrabasic and metamorphic rocks in high thermobaric regimes and some general problems of their change with depth (in Russian). Geofiz. journ., 5 (1983) 5, 35–45.

[3] Lebedev, T. S.; Polovinkin, B. V.; Korchin, V. A., et al.: Elastic properties of rocks from a deep borehole of the central Ukrainian Shield and laws of their changes under different thermobaric conditions of the Earth's crust (in Russian). — Geofiz. journ., 5 (1983) 4, 10–25.

[4] Lebedev, T. S.; Korchin, V. A.: Complex experimental studies of elastic properties of rocks under model PT conditions of the Earth's crust (in Russian). — Geofiz. sb. (1978) 84, 56–74.

[5] Lebedev, T. S.; Shapoval, V. I.; Korchin, V. A.: New data on longitudinal wave velocities in rocks at high thermodynamic parameters (in Russian). Geofiz. sb. (1972) 49, 9 to 27.

[6] Lebedev, T. S., et al.: Elastic properties of rocks at high pressures (in Russian). Kiev, Naukova Dumka (1972) 183 p.

[7] Lebedev, T. S.; Shapoval, V. I.; Korchin, V. A.: Programmed modelling of deep thermodynamic conditions in experimental studies of mineral matter elastic properties (in Russian). In: Geophysical studies of the Earth's crust and mantle of Ukraine. Kiev, Naukova Dumka (1975) 25–30.

[8] Lebedev, T. S.; Korchin, V. A.; Shapoval, V. I.: Simultaneous studies of longitudinal and transversal wave velocity in rocks and minerals in different experimental P-T regimes (in Russian). In: Studies of Earth's mineral matter physical properties at high thermobaric parameters (in Russian). Kiev, Naukova Dumka (1977) 18–33.

[9] Shermerger, T. D.: Theory of elasticity of microinhomogeneous media (in Russian). Nauka, 1979, 252 p.

[10] Clark, W. S.: Handbook on physical constants of rocks (in Russian). Mir (1969) 352 p.

[11] Krilov, V. D.: Use of Fourier-transformations of approximating functions in x-ray analysis of polycrystal structure (in Russian). Crystallography 8 (1963) 3, 486–489.

[12] Rusakov, A. A.: Roentgenography of metals (in Russian). Atomizdat (1977) 480 p.

[13] Iveronova, V. I.; Revkevich, G. P.: Theory of x-ray dispersion (in Russian). MGU (1982) 280 p.

[14] Vishnyakov, Y. D.: Recent methods of studies of deformated crystal structure (in Russian). Metallurgizdat (1975) 450 p.

Study of P-wave velocity and density of uzbekistan rock samples under high quasi-hydrostatic pressure up to 1.6 GPa

Muz. Kh. Bakiev, A. Kh. Ibragimov

The role of the laboratory study of rocks' physical properties at high thermodynamic conditions is growing in connection with the development of seismogeodynamics studying seismic effects of the different scale as one dynamic system and taking into account rocks' elastic and rheologica properties, the strength and the processes of deformation, cracking and fracture [1—4]. Such investigations are assuming ever greater importance for developing earthquake fault models and interpretation geophysical field observations data.

According to the results of the seismogeodynamic investigations the Uzbekistan territory is divided into three zones: Fergana zone (orogen of Tien Shan), Kizil-Kum zone (Turan plate) and Tashkent zone (transitional zone from Tien Shan orogen to the Turan plate). These zones are essentially different by seismic and geodynamic conditions, geologic-geophysical characteristics of the crust and its geological history [1].

Studied rocks were divided into four groups by their petrographic features: sedimentary-metamorphic group (hornstones and sandstones); carbonatic (limestones); porphyric (quartz-porphyries and granite-porphyries); and intrusive group (granites and grano-diorites). Samples were selected from the rock exposures as well as cores from the boreholes in order to study their physical properties. The rock samples' absolute and geological ages are given in Table 1. It is seen that the ages of rocks from different zones are approximately equal to each other.

Laboratory measurements of the P-wave velocity in the rock samples were performed in the cylinder-piston apparatus [5]. The propagation time of elastic waves was determined by the method of pulse sonic test at the frequency 1.0 MHz and high quasi-hydrostatic pressures up to 1.6 GPa. Lead was used as pressure transmitting medium.

Before studying rock elastic properties, the velocity and porosity, the density also has been determined at atmospheric pressure. The full chemical and petrographic analysis was made on the rock samples.

Carbonatic rocks. P-wave velocity and density in limestone samples were measured at pressures up to 1.6 GPa. The velocity varies with pressure identically for all three zones in the initial pressure range (Fig. 1); but the value of the velocity is higher in the rock samples from Kizil-Kum region and reaches 6.45 km/s. The samples of the transitional and orogenic zones are characterized by velocity values of 6.38 and 6.30 km/s, respectively. As the pressure increases to 1.0 GPa average values of the velocity

Table 1. Geological and absolute age (million years) of the rocks from Uzbekistan seismic zones

Rocks	Geological and absolute age (10^6 years)		
	Fergana zone	Tashkent zone	Kizil-Kum zone
Hornstones	O–S	O–S	O–S
Limestones	D_2–C_1	D_2–C_1	D_2–C_1
Granites	C_3 (280)	C_3 (180)	P_1 (260)
Granodiorites	C_2 (290)	C_2 (290)	C_3–P_1 (275–260)
Quartz-porphyries	P–T (260–240)	P–T (260–240)	S–D (360)
Granite-porphyries	C_3–P_1 (280–270)	C_3–P_1 (280–270)	P_2 (250)

Legend: O — Ordovician period; S — Silurian; D_2 — Middle Devonian; C_2 — Middle Carboniferous; C_1 — Upper Carboniferous; C_3 — Lower Carboniferous period; P — Permian; P_1 — Lower Permian; P_2 — Upper Permian; T — Triassic.

in the Kizil-Kum and Tashkent rock samples alter insignificantly, while in the Fergana limestones V_P begins to decrease to 6.12 km/s in the pressure range from 0.4 to 1.0 GPa. The pressure has been increased further, the velocity falls down to 4.74 km/s at the pressure of 1.6 GPa, as the result of the polymorphous changes of calcite into aragonite [6]. The average values of density vary in the same succession as the P-wave velocity: the highest density has been received for the Kizil-Kum rocks and the lowest one is for the Fergana rocks.

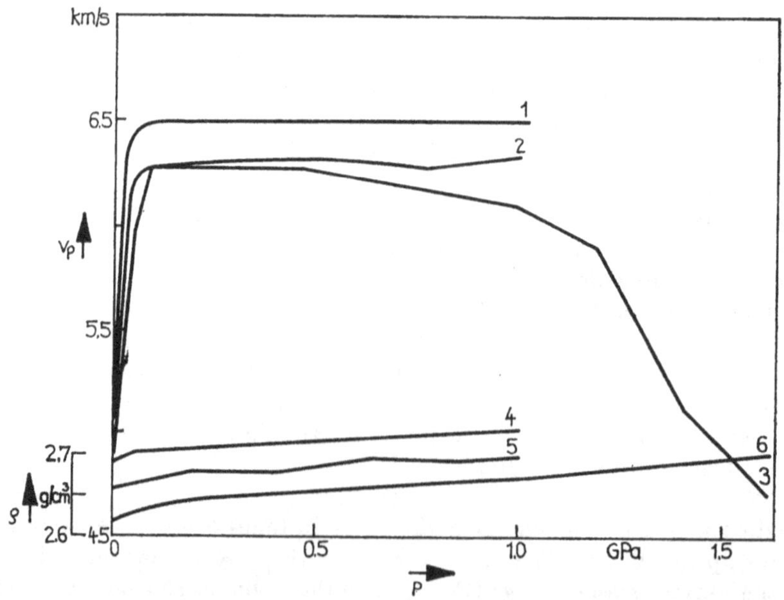

Fig. 1. Average values of the physical properties of limestone samples: *1–3* — P-wave velocity; *4–6* — densities; *1, 4* — for the Kizil-Kum zone; *2, 5* — for the Tashkent zone; *3, 6* — for the Fergana zone.

Sedimentary-metamorphic rocks. The experiments have shown that the rock samples from the Fergana, Tashkent and Kizil-Kum zones are characterized by the velocity values in the range of 4.35—5.54; 5.60—5.69 and 5.10—5.71 km/s respectively at atmosphere pressure. As the pressure reaches 1.0 GPa the velocity values are equal to 6.17—6.87; 7.23—7.61 and 5.69—6.50 km/s respectively. So, the P-wave velocities at atmosphere pressure are almost the same for all three regions. But with the pressure application it varies for different zones. For example, the velocities in the hornstones of Tashkent zone have increased by 2.0 km/s in the above stated pressure range, while for the Kizil-Kum samples they also do by the 0.79 km/s in the same pressure range.

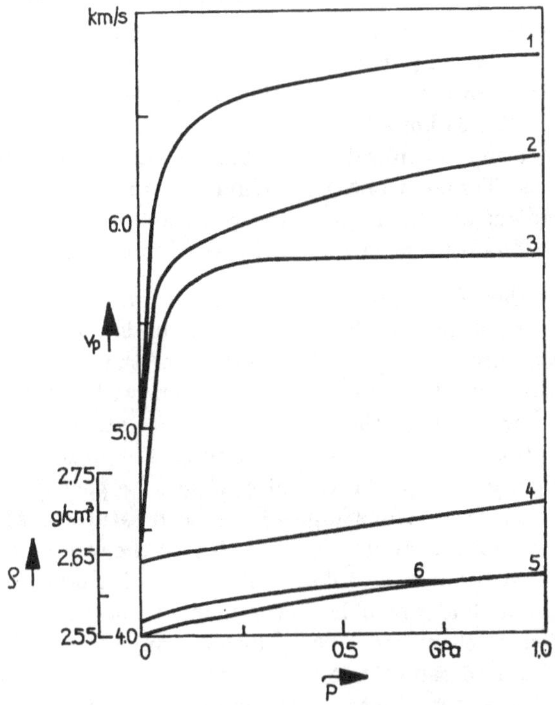

Fig. 2. Average values of the physical properties of granite samples. The notes are the same as for Fig. 1.

Porphyric rocks. The samples of quartz-porphyries and granite-porphyries were studied. The density of the rocks of the separate zones differs significantly in spite of the similar mineral composition. It can be explained by significant secondary changes in rocks of Tashkent zone, which have the lowest density — 2.2 g/sm³ at atmosphere pressure. The highest velocities have been obtained for the Fergana region quartz-porphyries and the lowest ones for the samples of the Tashkent zone.

Intrusive rocks. Most earthquakes of Uzbekistan are located at the deph corresponding to the bedding of granite-diorite rocks [11, 12]. Average velocity values at atmosphere pressure are in the range from 4.45 to 4.85 km/s in granites from all three zones: the highest values correspond to the Tashkent rocks and minimum one does to the Fergana samples (Fig. 2). The curves of the density mean values are drawn in the same succession and vary almost monotonous with pressure. Maximum velocity values at the pressure of 1.0 GPa are shown by the Kizil-Kum rocks and minimum ones — by the granites of the Tashkent zone.

It is known that elastic characteristics of rocks at the atmosphere pressure are defined in general by the intergranular porosity and cracks in the rock. The pressure application results in sharp increase of the V_P value, which is connected with the closure of most cracks and improvement of acoustic coupling between grains of the rock-forming minerals [5]. The degree of cracking is dependent greatly on the weathering of the tectonic stresses, effected on the rock for a long time in geological scale. Hence, the gradient of the P-wave velocity due to the pressure increase can be used as the characteristic of the tectonic situation in places the rock samples have been selected. Thus, for example, the average value of the velocity increment ΔV_P in the pressure range from the normal to 0.1 GPa is 1.30 km/s for the granites from Fergana zone. It is equal to 0.91 and 1.67 km/s for the rocks from Tashkent and Kizil-Kum zones, respectively. The velocity mean values are 0.86; 0.64 and 1.90 km/s for the granodiorites from Fergana, Tashkent and Kizil-Kum zones respectively. In both cases the highest velocity gradient due to the pressure is characterized for the granites of Kizil-Kum and the lowest one is for the rocks of the Tashkent zone.

Influence of Saturation. It is known that under natural conditions rocks are in interaction with fluids. The degree of the influence of fluids on an elastic wave velocity has not been studied sufficiently yet. The presence of the fluid phase in a pore space raises in general the velocity values in comparison with unsaturated rocks under the condition that the value of the pore pressure is rather small and that there is no chemical interaction between them. To study the influence of the saturation on the P-wave velocity change the samples of granite-porphyries, granites, granodiorites, etc. have been used. The samples have been saturated after the evacuation. The percentage of the saturation was 0.29—0.86% of the sample volume. The experiments have shown that the P-wave velocity of the rock including minerals to be interacted with a saturated fluid would be lower than in unsaturated rock samples. One should note the different character of the velocity change due to the pressure for some saturated and unsaturated samples of some types of rocks. So, some decrease of the velocity has been observed in the saturated granodiorites from the Tashkent zone in the initial pressure range (0.1—0.2 GPa), while it has not been observed for the unsaturated samples. The further pressure increase up to 1.4 GPa is followed by the velocity increase both in the saturated and unsaturated samples, so the velocity-pressure curves are almost parallel to each other.

Influence of mineral composition. The dependence of the P-wave velocity at high pressures on mineral content in samples of intrusive rocks have been obtained. It has been defined that the largest influence on general elastic properties of rocks belongs to biotite: insignificant rise of the biotite content in the rock (3—6%) results in the velocity increase by 0.4—0.8 km/s; the effect of the quartz or plagioclase content is not so strong. A clear correlation has not been obtained also for the potassic feldspar (correlation coefficient is in the range 0.20—0.38 under different pressures).

Conclusions. It is known [1], that the territory of the Turan plate is characterized by the more consolidated Earth crust, which is 5—10 km thinner, and the elastic wave velocity is 0.1 km/s higher at average than in Tien Shan orogen and in the transitional zone from the orogen to the plate [1, 8]. The results of the laboratory study of the elastic properties of rocks, given in the present paper, confirm the data of field observations and show that the rocks of Turan plate of the same type as the rocks of the Tien Shan orogen and transitional zone are charaterized by more high velocity values.

References

[1] Ulomov, V. I.: Dynamics of the crust in Central Asia and earthquake prediction (in Russian). Tashkent: FAN. (1974) 214 p.

[2] Ulomov, V. I.; Bykovtsev, A. S.; Flenova, M. G.: On methodological basis of the seismogeodynamics. Abstracts (in Russian) of the ICSEE Ac. Sci. USSR. Frunze: ILIM. (1981).

[3] Ulomov, V. I.: Tectonics of the lithospheric plates and seismogeodynamics (in Russian). In: Experimental seismology in Uzbekistan. Tashkent: FAN. (1983).

[4] Mavlyanov, G. A.; Ulomov, V. I.: Searches for the earthquakes' precursors in Uzbekistan (in Russian). In: Searches for the earthquakes' precursors. Tashkent: FAN. (1976) 25—37.

[5] Volarovich, M. P.; Bayuk, E. I.; Levykin, A. I.; Tomashevskaya, I. S.: Physico-mechanical properties of rocks and minerals at high pressures and temperatures (in Russian). Moscow: NAUKA (1974) 221 p.

[6] Volarovich, M. P.; Bayuk, E. I.; Efimova, G. A.: Elastic properties of minerals at high pressures (in Russian). Moscow: NAUKA. (1975) 131 p.

[7] Bakiev, Muz. Kh.; Ulomov, V. I.; Bakiev, M. Kh.; et al.: Comparison of physical properties of rocks from regions of Uzbekistan with different seismicity (in Russian). In: Earthquake prediction, Dushanbe—Moscow: DONISH. (1983—1984) 4, 232—243.

[8] Bakiev, Muz. Kh.; Kireenkova, S. M.; Ibragimov, A. Kh.; Akhundjanov, R.: Elastic properties of rock samples from different seismogenous zones at high quasi-hydrostatic pressures up to 18 Kbars. (in Russian). In: Seismogeodynamics of the transitional zone from Tien-Shan orogen to Turan plate. Tashkent: FAN. (1986) 129—135.

[9] Bakiev, Muz. Kh.; Kireenkova, S. M.; Ibragimov, A. Kh.: Study of the P-wave velocities and densities of Uzbekistan rock samples in high thermodynamic conditions (in Russian). Izvestiya Ak. Nauk SSSR, ser. Physics of the Earth. (1984) 11, 110—118.

[10] Bakiev, Muz. Kh.; Ibragimov, A. Kh.; Bakiev, M. Kh.: Effects of mineral composition on the elastic properties of Uzbekistan rocks (in Russian). In: Physical properties of rocks at high pressures and temperatures. Erevan. (1985) 22—28.

[11] Butovskaya, E. M.; Djamalov, D. B.; Siradgitdinov, V. B.: Foci' distribution of the earthquakes by the depth in East Uzbekistan and its mountain margins in comparison with the crust layers (in Russian). Uzb. Geol. Journ. (1980) 6, 3—5.

[12] Yudakhin, F. N.: Geophysical fields, deep structure and seismicity of Tien Shan (in Russian). Frunze: ILIM. (1983) 248 p.

Influence of extreme conditions on the elastic anisotropy of rocks

E. I. Bayuk, F. M. Levitova

Information about seismic anisotropy makes it possible to define more exactly the structure and substantial composition of the Earth's interior. Investigations carried out by the method of refracted waves and the depth seismic sounding have revealed a velocity anisotropy of longitudinal and transversal waves in the Earth's crust lower part and the upper mantle, as well as of Love and Rayleigh waves in the areas of the Pacific and Atlantic Oceans. Seismic anisotropy was also found under the continents [1, 2].

There exist several hypotheses for explaining the nature of elastic anisotropy, one of which relates the appearance of velocity anisotropy to the preferred orientation of rock-forming minerals [3, 4].

This work is dedicated to the laboratory studies of the change in the velocity of elastic waves in rocks depending on the preferred orientation of rock-forming minerals and the effect up to 1.5 GPa pressure and up to 600 °C temperatures upon the elastic anisotropy factor.

Chosen as samples for investigation were such olivine-bearing rocks as perodotites, olivinites and dunites. The reason for this choice is first that these rocks are supposed to be characteristic for the upper mantle and second because the olivine mineral possesses a considerable elastic anisotropy and can be in a definite way orientated in the rocks.

Investigations were conducted in the following succession. At first, the orientation of olivine grains in the rocks was determined. Then the velocities of the longitudinal and transversal elastic waves were measured in various directions in the samples to find out the connection between the velocity values and the rock internal structure. At last, the velocity of elastic waves was measured at high pressures and temperatures in cylindrical samples definitely cut-out in compliance with the preferred orientation of olivine grains to study the influence of thermodynamic conditions upon the elastic anisotropy of rocks.

Equipment and Methods of Investigations

The velocity of longitudinal and transversal waves was measured by the ultrasonic pulse method at a frequency of 1.8 to 2 MHz. Ultrasonic equipment comprizing Model G5-15 generator, USh-10 amplifier, C1-13A oscillograph and piezoelectric transducers made of ZTS-19 ceramics and piezoquartz was used.

Experiments at high pressures up to 1.5 GPa were performed in an installation of the cylinder-piston type, with solid matter (lead) as the pressure-transfer agent [5], or pyrophyllite for the case of experiments at high temperatures [6]. Cylindrical samples, 2 cm in length and 1.7 cm in diameter, were used for the experiments. Two series of experiments at high pressures were conducted. In one of the series we studied the velocity anisotropy of longitudinal and transversal waves at high pressures, while the other series was aimed at studying the velocity anisotropy of only the longitudinal waves, with simultaneous influence of pressure and temperature. This resulted in the application of two methods and caused a difference in the mounting of the samples and the piezoelectric transducers.

Table 1. Velocity of Elastic Waves in Olivine Crystals (km/s) [8]

Direction of propagation in wave	Direction of dislocation in wave	Forsterite $\varrho = 3.22$ g/cm³	Olivine $\varrho = 3.31$ g/cm³
[100]	[100]	10.9	9.89
[100]	[010]	5.01	4.89
[100]	[001]	5.02	4.87
[010]	[010]	7.87	7.73
[010]	[100]	5.01	4.89
[010]	[001]	4.52	4.42
[001]	[001]	8.54	4.83
[001]	[100]	5.02	4.87
[001]	[010]	4.52	4.42

To ensure simultaneous measurement of the velocities of longitudinal and transversal waves, in addition to the pair of longitudinal wave piezoelectric transducers, the anvil accommodated a pair of transversal wave transducers. The signals from the two pair of transducers were received by two similar sets of ultrasonic equipment [7].

As known, not only the longitudinal wave velocity can be measured in the main crystollographic lines in the olivine monocrystal, but also two transversal wave velocities differing in the dislocation in the wave. This is seen from Table 1 which presents the experimental data for olivine and forsterite, taken from [8]. No difference in the velocities of transversal waves in the [100] direction was noticed, whereas in the [010] and [001] directions it was quite considerable (up to 11 per cent) for orthogonal directions of dislocation in the wave. In connection with the foregoing, the transversal wave velocities for the selected rocks were measured in each sample in two successive experiments, the relative position of the piezoelectric transducers being 90°.

A composite copper-graphite short-circuited heater which enables to maintain quite a stable temperature pattern was housed inside the chamber during the high-temperature experiments. The temperature of rock samples was measured with the aid of a Chromel-Alumel thermocouple. Three ultrasonic piezoelectric transducers were used for measuring the longitudinal wave velocities. One of them (the ZTS ceramics) was arranged in the lower temperature zone of the anvil and served as a radiator, the other two (piezoquartz ones) were mounted right on the sample and fulfilled the functions of a receiver. The time of the ultrasound passage through the sample was determined by the difference between the times of the first arrival in both receivers. First the pressure was increased in the experiments to a value of 0.5, 1.0, 1.5 GPa, then the temperature was increased up to 600 °C, with the pressure being constant. Such a method allowed to avoid cracking of the heated samples caused by the difference in expansion of the multiphase medium, to which the rock pertains. High pressure prevents formation of joints.

Results

The diagrams obtained during the microstructural study of the sections show definite regularities in the spatial distribution of the axes of olivine crystals. The orientation of the optical axes of olivine grains which correspond to the [100], [010], [001] crystallographic axes was determined. Anisotropic structure of the rock is manifested by the fact that the points, corresponding to the measurements of similar axes in different grains

[100] [010] [001]

Fig. 1. Diagram of orientation of axes of 150 olivine grains for peridotite. Isolines for 1, 2, 3, 4, 5, 6 (%).

of rock minerals, are located close to each other and form an accumulation of points. The more perfect the orientation, the higher the density of location of the points. The measurement results are introduced into the diagram for three axes and thus three diagrams are obtained.

Along with obvious distinctions, there is also much in common in the studied microstructural diagrams of ultrabasic rocks. Thus, for example, all the diagrams revealed orthogonal position of axes maxima (Fig. 1). In some cases complete or partial bands along one or two axes were observed. Sections of approximately similar intensity are possible within the bands. Predominance of some preferred orientation was found in each sample. Most of the rocks studied possess strong maxima along all the three olivine axes, i.e. they show a type of symmetry close to the rhombic one. The velocities of longitudinal and transversal waves were measured at atmospheric pressure in various directions in the rock blocks [9]. Diagrams showing the distribution of velocities in the rock were plotted on the basis of the above measurements. The comparison between the diagrams of orientation of olivine grains and the diagrams showing the distribution of velocities in the rocks made it possible to state that the maximum velocities of elastic waves correspond to the maximum of axes [100], the minimum velocities correspond to the maximum of axes [010], while the average velocities correspond to the maximum of axes [001] (Fig. 2).

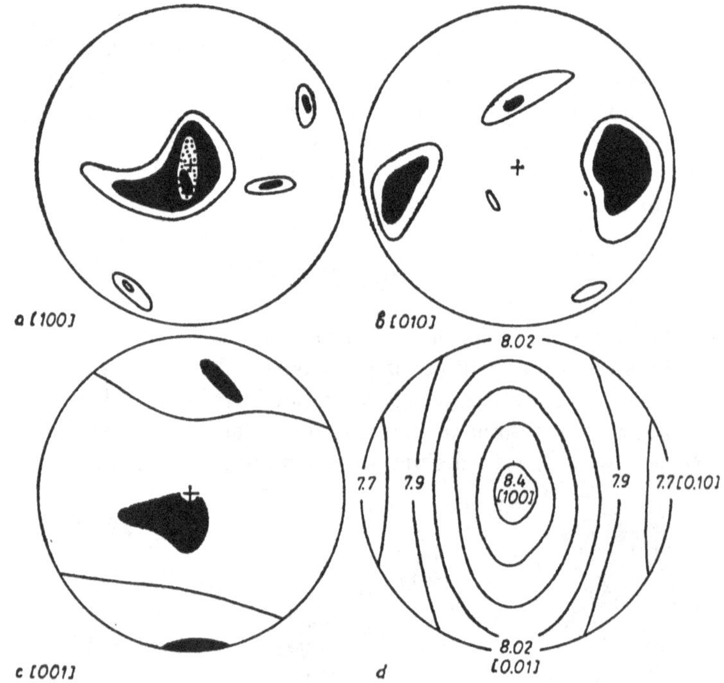

Fig. 2. Diagram of orientation of axes of 100 olivine grains and diagram of spatial distribution of longitudinal waves velocities for peridotite (Number — experimental data of velocities of longitudinal waves in km/s).

High values of the velocity anisotropy factor were observed: (a) for structures with a strong maximum of axes [100] and with partial bands for axes [010] and [001]; (b) for structures with a strong olivine maximum of axes [010] and emergence bands for axes [100] and [001]. Low values of the anisotropy factor were obtained for structures having a strong maximum of axes [001] and emergence bands [100] and [010]. Samples with weak olivine orientation possess low variations of velocities in different directions. The preferred orientation of the axes of olivine grains in the investigated samples made up 15 to 35 per cent of the total content of olivine in the rock.

The results of the experimental investigation of the elastic wave velocities in olivine-bearing rock samples at high pressures are presented in Table 2. The Table specifies the directions along which the velocities and dislocations in the wave were measured. We did not succeed in determining the velocity of transversal waves for all the rocks. It is seen, however, that the same as for the monocrystal, the velocities of transversal waves in the [100] direction have a slight difference, whereas in the other directions they differ considerably.

The high-pressure experiments have proved that the directions of minimum and maximum velocities in olivine-bearing rocks do not alter at pressures up to 1.5 GPa as well. These data add to the pattern of spatial distribution of velocities; the maxima and minima positions remain unchanged, on the whole. Variation of pressure causes a change only in the configuration of the velocity isolines, due to a considerable increase of the velocity in the direction of minimum velocities. The rate of velocity in-

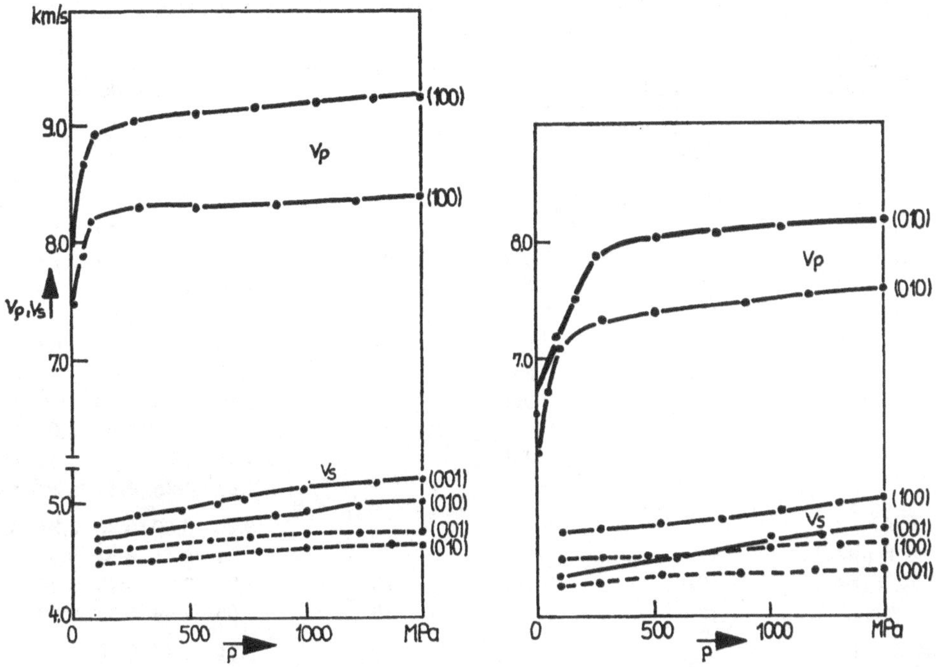

Fig. 3. Alteration of longitudinal waves velocities vs. pressure for olivinite (continuous line) and peridotite (dotted line) [100] and [010] — crystallographic trends.

crease at a pressure rise of up to 0.4 GPa is 6 to 23 per cent. Such a sharp velocity increase is indicative of a structural heterogeneity, i.e. of high porosity, in these particular samples. Some of the samples studied possess an effective porosity equal to 1.5—2 per cent. The most sharp velocity increase is noticed in olivinites up to a pressure of 0.1—0.2 GPa; in peridotites the increase is lower and smoother. Within the interval of 0.5—1.5 GPa the increase in the velocities of elastic waves makes up 1 to 3 per cent, since due to the effect of closing of cracks and microporosity, and the velocity reaches the values determined by the mineral composition of the rocks.

Fig. 3 illustrates the velocity curves of longitudinal and transversal waves with variation of pressure for olivinite (solid lines) and peridotite (dotted lines) samples cut out in the [100] direction (left) and [010] (right). As seen, the maximum values of velocities of transversal wave correspond to the maximum velocities of longitudinal waves in the [100] direction, while the minimum values of velocities of transversal wave correspond to the minimum values of velocities of longitudinal waves in the [010] direction. The difference in the velocities of transversal waves for one direction versus the dislocation made up 3 to 6 per cent for different samples.

In the high-pressure experiments not only the velocity of elastic wave, but also the density was determined (Table 2). At atmospheric pressure almost all the samples of ultrabasic rocks possess a density exceeding 3.00 g/cm³, save for the dunite sample, whose density at atmospheric pressure is 2.97 g/cm³, due to serpentinization of the rock. With the increase of pressure up to 1.5 GPa the density in all the tested samples increases by a comparatively low value within 1.5—3.7 per cent.

Table 2. Velocities of Elastic Waves (V) and Density (ϱ) in Oriented Speciments of Ultrabasic Rocks under High Pressures

Rock Location Mineral composition %	Direction of propagation of wave	Direction of dislocation in wave	Orientation %	ϱ (g/cm³), V (km/s) at pressures (GPa) 0,001	0,1	0,5	1,0	1,5
1	2	3	4	5	6	7	8	9
Peridotite-1		ϱ		3.28	3.29	3.32	3.33	3.33
Peninsula Kola,	[100]	[100]	20	7.50	8.20	8.28	8.33	8.40
Ol-60, Pyr-35,	[100]	[010]		4.40	4.50	4.56	4.60	4.62
Mag-5	[100]	[001]		4.55	4.64	4.68	4.74	4.76
	[010]	[010]	18	6.10	7.20	7.50	7.65	7.70
	[010]	[100]		4.39	5.52	4.58	4.60	4.63
	[010]	[001]		4.10	4.25	4.35	4.38	4.40
	[001]	[001]	20	6.80	7.78	7.94	8.00	8.02
	[001]	[100]		4.55	4.64	4.68	4.74	4.76
	[001]	[010]		4.10	4.25	4.35	4.38	4.40
Peridotite-2		ϱ		3.29	3.30	3.33	3.34	3.34
Kola Peninsula	[100]	[100]	22	7.84	8.48	8.65	8.70	8.74
Ol-52, Pyr-40,	[010]	[010]	23	7.20	7.70	8.20	8.28	8.32
Pl-5, Mag-3	[001]	[001]	20	7.65	7.98	8.36	8.50	8.56
Olivinite-1		ϱ		3.34	3.45	3.46	3.46	3.46
Kola Peninsula	[100]	[100]	35	8.00	8.95	9.14	9.24	9.28
Ol-90, Pyr-5,	[100]	[010]		4.65	4.82	4.95	4.98	5.02
Mag-5	[100]	[001]		4.70	4.84	4.96	5.09	5.16
	[010]	[010]	25	6.70	7.25	8.06	8.18	8.20
	[010]	[100]		4.55	4.70	4.80	4.90	5.00
	[010]	[001]		4.20	4.35	4.50	4.65	4.68
	[001]	[001]	30	7.42	7.75	8.16	8.28	8.33
	[001]	[100]		4.70	4.84	4.96	5.09	5.16
	[001]	[010]		4.20	4.35	4.50	4.65	4.68
Olivinite-2		ϱ		3.31	3.32	3.34	3.35	3.36
Siberia	[100]	[100]	15	7.18	7.70	8.08	8.18	8.20
Ol-97,	[001]	[001]	15	6.50	7.00	7.80	7.90	7.95
Mag-5	[001]	[001]	15	6.50	7.00	7.80	7.90	7.95
Olivinite-3		ϱ		3.20	3.21	3.21	3.22	3.23
Siberia	[100]	[100]	35	6.98	7.50	7.86	7.90	7.91
Ol-85, Mag-15	[010]	[010]	35	5.94	6.50	6.85	6.98	7.00
Olivinite-4		ϱ		3.21	3.22	3.24	3.25	3.26
Siberia	[100]	[100]	30	6.34	7.60	7.96	8.06	8.10
Ol-85, Pyr-10, Mag-5	[010]	[010]	30	5.50	6.72	7.14	7.22	7.26
Dunite		ϱ		2.97	2.98	3.03	3.05	3.05
Russian Platform	[100]	[100]	27	6.60	6.80	7.00	7.08	7.12
Ol-60, Pyr-7,	[010]	[010]	27	5.56	5.80	6.00	6.20	6.28
Serp-30, Chr-3	[001]	[001]	27	6.00	6.45	6.92	6.94	7.00

Note: Ol — olivine, Pyr — pyroxene, Pl — plagioclas, Serp — serpentine, Mag — magnetite, Chr — chromite.

Table 3. Values of Velocity Anisotropy Factor (A) in Ultrabasic Rocks
1 — Direction [100]; 2 — Direction [010]; 3 — Direction [001].

Rock	A	Wave type	A (%) at a pressure of GPa				
			0.001	0.1	0.5	1.0	1.5
Peridotite 1	A_{12}	P	19.67	13.89	10.40	10.03	9.09
	A_{12}	S	8.29	9.18	8.59	8.22	8.18
	A_{13}	P	7.35	5.40	4.28	4.13	4.07
	A_{23}	P	11.48	8.06	6.94	5.96	4.73
Peridotite 2	A_{12}	P	8.89	10.13	5.49	5.07	5.05
	A_{13}	P	2.48	6.27	3.47	3.35	3.10
	A_{23}	P	6.25	3.64	3.00	2.67	2.88
Olivinite 1	A_{12}	P	19.40	23.45	13.40	13.00	13.17
	A_{12}	S	11.91	11.49	10.22	10.23	10.25
	A_{13}	P	7.82	15.48	12.00	11.59	11.40
	A_{23}	P	10.75	6.89	1.24	1.22	1.58
Olivinite 2	A_{13}	P	10.46	10.00	3.59	3.54	3.14
Olivinite 3	A_{12}	P	17.51	15.38	14.74	13.80	13.00
Olivinite 4	A_{12}	P	15.27	13.10	11.76	11.63	11.57
Dunite	A_{12}	P	18.70	17.24	16.70	14.20	13.30
	A_{13}	P	9.00	6.30	1.10	2.02	1.70
	A_{23}	P	7.91	7.75	13.30	11.90	11.50

The data obtained by measuring the velocities of elastic wave with variation of pressure served as a basis for determining the change of the velocity anisotropy factor. For all the tested rocks and all directions there was noticed a reduction of the velocity anisotropy factor with increasing pressure (Table 3). Maximum reduction at a pressure up to 0.2—0.3 GPa is due to the closing of joints and pores. At high pressures the variation of the anisotropy is, as a rule, insignificant. The difference in the velocities for directions [100] and [010], or [100] and [001] becomes herein smaller, while for the [010] and [001] directions it gets somewhat larger.

The effect of temperature upon the velocity of longitudinal waves in olivine-bearing rocks at high pressures will be found in Table 4 and Fig. 4. As seen, under the effect of temperatures ranging from room temperature to 600 °C the velocity varies by approximately 3 per cent; at a pressure of 0.5 GPa the variation is somewhat higher than at 1.0 GPa and 1.5 GPa. The nature of the velocity change with variation of temperature does not actually depend on the direction of wave propagation.

The graph presented in Fig. 4 shows the variation of the anisotropy of longitudinal wave velocity versus pressure (left) and temperature (right) for olivinite (*1*) and peridotite (*2*) samples. As seen from the graph, at a pressure up to 0.5 GPa the isotherms have a steeper slope, then they flatten out and run nearly parallel to each other. With a rise of pressure the velocity anisotropy factor decreases, whereas with a rise of temperature under high pressure conditions it does not in fact change, i.e. the velocity isobars are actually parallel.

Table 4. Velocities of Longitudinal Waves in Oriented Samples of Ultrabasic Rocks at Various Temperatures and Pressures

Rock	Orientation of axes	Pressure, GPa	Temperature, °C				
			20	100	200	400	600
Peridotite 1	[100]	0.5	8.28	8.23	8.17	8.05	7.93
		1.0	8.33	8.29	8.24	8.13	8.03
		1.5	8.40	8.36	8.31	8.20	8.10
	[001]	0.5	7.94	7.89	7.83	7.71	7.59
		1.0	8.00	7.96	7.91	7.80	7.70
		1.5	8.02	7.98	7.93	7.82	7.72
	[010]	0.5	7.50	7.45	7.39	7.27	7.15
		1.0	7.65	7.61	7.56	7.45	7.35
		1.5	7.70	7.66	7.61	7.50	7.40
Olivinite	[100]	0.5	8.08	8.02	7.95	7.81	7.67
		1.0	8.18	8.14	8.08	7.98	7.87
		1.5	8.20	8.16	8.11	8.01	7.91
	[001]	0.5	7.80	7.74	7.67	7.53	7.38
		1.0	7.90	7.86	7.81	7.69	7.58
		1.5	7.95	7.91	7.86	7.77	7.66
Dunite	[100]	1.0	7.08	7.06	7.04	6.99	6.94
		1.5	7.12	7.10	7.08	7.03	6.98
	[010]	1.0	6.20	6.18	6.15	6.10	6.05
		1.5	6.28	6.26	6.24	6.19	6.14

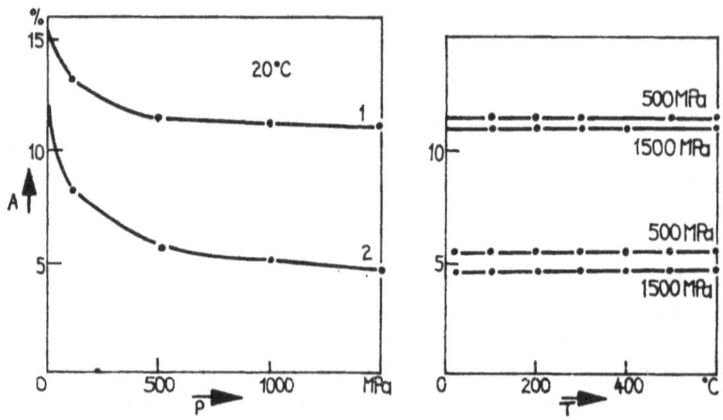

Fig. 4. Isotherms (left) and isobars (right) of anisotropy coefficients of velocity of longitudinal waves: *1* — for olivinite, *2* — for peridotite.

Geophysical Discussion of the Results

The investigations have shown that the value of the velocity anisotropy factor is influenced by the degree of orientation and the content of rock-forming minerals. For studying the samples of ultrabasic rocks, Fig. 5 shows the velocity of longitudinal wave factor related to the quantity of oriented olivine grains. The correlative dependence between the velocity anisotropy factor and the quantity of oriented olivine is assumed

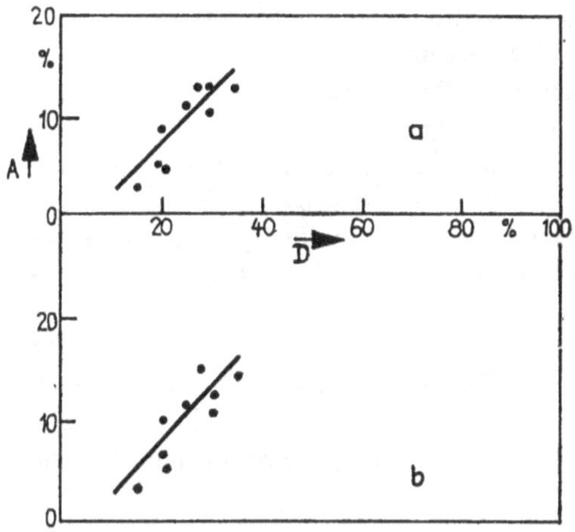

Fig. 5. Alteration of elastic anisotropy coefficients vs. Number of oriented grains of olivine in olivine-bearing rocks under high pressures. Points — experimental data of velocities of longitudinal waves (km/s) under 1.5 GPa — a, under 0.5 GPa — b.

to be a linear one, with the correlation coefficient equal to 0.91. With increased orientation of olivine grains the velocity anisotropy factor at constant pressure increases as well. In the rocks investigated, the velocity anisotropy factor at a pressure of 1.5 GPa and a temperature of 600 °C is of rather a high value, viz. 8 to 13 per cent.

The theoretical data were compared to the experimental ones. The moduli of elasticity and the respective velocities of elastic waves were calculated on the basis of the mineral composition and the obtained preferred orientation diagrams, from which it is obvious that the rocks studied have a type of symmetry close to the orthorhombic one (Table 2). The moduli of elasticity, as well as the calculated and experimental data of velocities in the corresponding directions are listed in Table 5. The experimental data of elastic

Table 5. Estimated Values of Velocity Elasticity Moduli for a Rhombic Type of Symmetry, and Their Calculated ($V_{calc.}$) and Experimental ($V_{exp.}$) Velocities

Rock	Parameter	$C_{ii} \cdot 10^{-11}$ (dyn/cm²) and V (km/s)					
		C_{11}	C_{22}	C_{33}	C_{44}	C_{55}	C_{66}
Peridotite	C	24.40	21.88	22.98	7.01	7.32	7.26
	$V_{calc.}$	8.47	8.02	8.22	4.30	4.64	4.62
	$V_{exp.}$	8.25	7.45	7.90	4.30	4.68	4.58
Peridotite	C	24.28	21.28	22.34			
	$V_{calc.}$	8.41	7.88	8.07			
	$V_{exp.}$	8.62	8.00	8.20			
Olivinite	C	26.93	23.11	24.49	7.27	7.73	7.65
	$V_{calc.}$	8.86	8.21	8.49	4.59	4.74	4.72
	$V_{exp.}$	9.05	8.00	8.10	4.52	4.88	4.81
Dunite	C	19.36	15.96	17.33			
	$V_{calc.}$	7.81	7.09	7.39			
	$V_{exp.}$	7.00	6.00	6.92			

wave velocities were obtained at pressures of 300 to 400 MPa, since the velocity calculations are done with the assumption that the rock is monolithic, i.e. it is free of cracks, structural defects, etc. The comparison between the calculated and experimental data shows that they are nearly alike. The difference in the velocities is not above 3—4 per cent, except for the dunite sample, where the difference in the minimum direction reaches 18 per cent, this is caused by the defects in the rock investigated.

Thus, the accomplished experiments make it possible to draw the following conclusion. The velocity anisotropy of elastic waves in ultrabasic rocks depends on the preferred orientation of olivine grains. With oriented olivine content of up to 35 per cent, the velocity anisotropy factor in the rock may reach 13 per cent. The effect of high pressure and temperature does not considerably reduce the elastic anisotropy of olivine-bearing rocks. This serves as a reason for stating that the possibility of existence of upper mantle seismic anisotropy owing to the preferred orientation may be taken for granted.

References

[1] Geophys. J. Roy. Astron. Soc. **49** (1977) 1.

[2] Christensen, N.; Salisbury, M.: Seismic anisotropy in the oceanic upper mantle. Evidence from the Bay of Islands ophiolite complex. J. Geoph. Res. **84** (1979) B 12, 6849—6857.

[3] Hess, H.: Seismic anisotropy of the uppermost mantle under Oceans. Nature **203** (1964) 6, 629—631.

[4] Chesnokov, E. M.: Seismicheskaja anisotropy verchnei mantii Zemli. Nauka (1977) 144 p.

[5] Volarovich, M. P.; Bayuk, E. I.; Levitova, F. M.: Metod odnovremennogo ismerenija uprugich i plotnostnich svoistv gornich porod pri davlenii do 20 kbar. Gerl. Beitr. Geophys. **86** (1977) 1, 69—78.

[6] Volarovich, M. P.; Bayuk, E. I.; Shaginjan, G. Sh.: Issledovanie skorosti prodolnich voln v gornich porodach i mineralach pri davlenijach do 15 kbar i temperaturach do 600 °C. Isv. AN SSSR, Physika Zemli (1977) 7, 82—89.

[7] Safarov, I. B.; Volarovich, M. P.; Levitova, F. M.; Kireenkova, S. M.: Anisotropy skorosti poperechnich voln vollivinovich porodach pri davlenii do 15 kbar. Isv. AN SSSR, Physika Zemli (1981) 9, 38—43.

[8] Kumazawa, M.; Anderson, O.: Elastic moduli pressure derivatives and temperature derivatives of single-crystal olivine and single-crystal forsterite. J. Geoph. Res. **74** (1969) 5961—5972.

[9] Bayuk, E. I.; Volarovich, M. P.; Levitova, F. M.: Uprugaja anisotropija gornich porod pri visokich davlenijach. M., Nauka (1982) 169 p.

The comparative analysis of physical properties of alpinotype ultrabasalts in some geologic regions

Sh. A. Balakishiev, A. B. Gasanov

Basic-ultrabasitic formations characterizing the eugeosynclinal stages in the history of the Earth's evolution are assumed by many investigators to be the result of upper mantle matter transformation. The study of these formations will enlarge our concept of inner structure and content of lower and upper mantle. The nature of basic-ultrabasic formations is usually considered from the point of view of ophiolitic associations and it is considered to be rather the same for geosynclinal regimes and postgeosynclinal epochs in all the folded regions. And the distinctions are due to non-uniformity of mantle, geologic and structural position of massifs of the basic and ultrabasic rocks, conditions and degree of differentiation, etc.

In such a situation it is natural to think that the common nature of the above mentioned rocks presupposes that the values of physical parameters characterizing them are very close to each other.

To check these assumptions the physical properties of alpinotype ultrabasites of the Minor Caucasus, Sayany-Baikal and Koriak-Kamchatka folded regions have been identified. For comparison there have been used data of laboratory studies of peridotites, pyroxenites, dunites samples under the conditions of high pressures and temperatures of similar petrographic, petrochemical and mineral composition.

Before presenting the results of the conducted investigations we should dwell on the notions on alpinotype ultrabasites genesis in the mentioned regions. As it has been mentioned this group of rocks is usually confined to ophiolitic associations. There are several points of view on the genesis of these associations. We shall present three of them.

1. Magmatic hypothesis. The followers of this hypothesis [1, 2, 3], which appeared at the end of 19th and the beginning of the 20th century, hold the opinion on intrusive genesis of ophiolitic complex rocks. According to this point of view ophiolites reveal deep fractures along which in different geologic time ultrabasic and basic magmas have been intruded. The formation of these magmas have been accompanied by differentiation and postmagmatic activity. However a number of geologic factors from the very beginning contradicted the magmatic hypothesis. For example, the absence of high-temperature active contacts of ultrabasic rocks with the surrounding enclosing rock which can be explained by the effect of ultrabasic magma, i.e. the rocks, into which the intrusive complex is intruded are not known.

To avoid this contradiction the volcanic hypothesis of ophiolites genesis has been advanced. According to it [4] ophiolites are the huge stream of basaltic lava that had ran out on the sea bottom. The stream got cold from the surface and the further magma cooling occurred inside the "sack" where in the result of gravitational differentiation the rugged section of ophiolitic series appeared. Petrologists and geochemists [5] who deny genetic relationship of ultrabasic and basic rocks of ophiolitic section raise some objections to this hypothesis. The depth of ultrabasic rock crystallization ob banded complex and gabbro is also at issue: on one hand they should be crystallized under the conditions of constant high temperatures according to all the petrological estimations, i.e. at great depth; and on the other hand — they have the same section with lavas what points to a small depths of the magma pocket in which crystallization took place. Besides the mentioned points the volcanic hypothesis is met with other difficulties.

2. Mantle hypothesis. The followers of this hypothesis [6, 7], based on the recent data of geologic science and presenting a new theory of global tectonics and location of ophiolites in the earth's crust, believe that ultrabasites are the blocks of mantle or solidificated at great depth intrusions tectonically moved to upper layers of lithosphere. According to this hypothesis the rocks of gabbroid series are either the parts of "basaltic" layer evacuated with ultrabasites or later intrusions.

Many petrologists [5, 9] explaining the constant coexistence of ultrabasic and basic rocks assume gabbro to be metasomatic formations which have appeared from ultrabasites under the conditions of high temperatures, i.e. at the level of upper mantle or lower part of "basaltic" layer. Some investigators [10, 11] assume ultrabasites to be the secondary rocks which have appeared in the result of protoplanet matter transformation. In particular, the paper [11] presents the proofs of relict nature of serpentine in gabbroids, pyroxenites and harzburgites. It is noted that gabbroization of serpentinites is the complex of intercorrelated processes: pyroxenization, feldspathisation, dehydration and recrystallization and the formation of ultrabasic complexes takes place, as the authors believe, in two stages which differ in time and in the conditions of development. During the first stage the solid dispared ultramafitic matter chemically similar to serpentinite has been formed. During the further stage this matter in the result of progressive transformation processes accompanied by recrystallization, dehydration and metasomatosis is subject to a number of metamorphoses with the formation of different types of serpentinites, dunites, harzburgites, olivine and serpentinic gabbro, amphibolites, hornblende gabbro, etc.

The complex character of petrogenic processes points to the fact that the formation of ophiolitic association formation embraces the longer period than the formation of intrusions.

3. Mantle-magmatic hypothesis. Mantle-magmatic hypothesis appeared in the result of detailed investigations [12] of ophiolitic series. It has been revealed that dunite-harzburgitic part of the section including lherzolites had undergone a number of complex and repeated deformations. Basaltoids and banded complexes and free of indications of such deformations. This points to a temporal gap between the formation of rocks of lherzolite-dunite-harzburgite and banded complexes. In this case dunites and harzburgites can represent solid residuum which has appeared during basalt melting from the mantle matter-pyrolite, and layered complex and gabbro are the products of basaltic magma differentiation intruded into this residuum. This is proved by the fact that the layered complec occurs not only above than the visible part of dunite-harzburgite complec but inside it as well [7]. The followers of this hypothesis believe

that such section of ophiolitic association appears within mid-oceanic ridge where along its axial fracture pyrolitic asthenolith intrudes what is accompanied by basaltic magma melting and origination of dunite-harzburgitic complex rocks in the form of solid residuum. Though this hypothesis, reflecting pure mantle and classical intrusive hypothesis, is not ideal, it explains many facts which could not be explained by earlier hypotheses. It proves the assumption of formation of layer "2" and "3" of the modern oceans crust in the zone of mid-oceanic ridges [12—17].

To estimate the significance of this conclusion for the problem of tectonic setting and the history of ophiolitic associations appearance we'll consider the section of oceanic crust obtained on the basis of data of drilling and dredging of oceanic bed on the slopes of reef zones, mid-oceanic ridges, in the flanges of gorges, etc. [14, 15].

From the top to the bottom oceanic crust is differentiated into three layers, the first one — the layer "1" or the layer of consolidated sediments is composed of carbonaceous and organogenic muds (including radiolarites) and other pelagic sediments.

The layer "2" or the layer of unconsolidated sediments is composed of tholeiitic and alkali basalts.

The composition of the layer "3" or "basaltic" layer of the oceanic crust at present is under discussion. Some authors [16] think that it is composed of serpentinites, others [14, 15] — of different metabasites (gabbro, amphibolites and greenstone rocks). Almost the same opinion based on the study of ophiolitic associations of the continents is expressed in [7] as well.

There is a third point of view on the composition of layer "3" of the oceans according to which [17] "basaltic" layer of the oceanic crust is composed of the mixture of hyperbasites, different gabbroids and metamorphic rocks. Thus, depending on the predominance of a particular type of rock the layer section can change from place to place. Many authors think that serpentinitic bodies penetrated into this layer as cold intrusions of upper mantle [6]. Below the layer "3" upper mantle occurs composed of rocks of ultrabasic composition, as most of the investigators think [12, 17].

The presented section of oceanic crust is very similar to the section of ophiolitic complex composed also from the bottom upwards of the layer of serpentinized ultrabasalts; layer of gabbroids and different metabasites; and at last of the layer of effusions of basaltic composition and pelagic sediments. This led to the conclusion that rocks of ophiolitic association occurring at present inside the folded regions, characterized by continental type of the Earth's crust structure, are the relics of the structures of oceanic type crust which sometimes occurred near the folded zones [8, 13].

The above presented points of view on genesis of ophiolitic associations reveal the complex character of this problem. It is necessary to continue data accumulation for a particular hypothesis. In this respect it is advisable to use geophysical methods and, in particular, the methods of laboratory model test of deep environment to reconstruct tectonic environment of the formation of ophiolitic associations rocks in particular geologic regions.

Based on above mentioned the samples of ultrabasites of the Minor Caucasus, Sayany-Baikal and Koriak-Kamchatka folded regions, which had been investigated earlier under normal conditions, underwent thermodynamic effect in the devices of high pessure. The changes of physical parameters in samples as function of pressure and temperature have been registrated (Table 1). The obtained data allow to distinguish the regions of change of the physical characteristics values under different thermodynamic

Table 1. Physical properties of ultrabasites in particular geological regions under normal and high pressures

Location of sampling	Rock type	NN samples	Density (g/cm³)	Longitudinal wave velocity (km/s) at pressures (MPa)				
				0.1	100	500	1100	1500
1	2	3	4	5	6	7	8	9
Lesser Caucasus	Dunite serpentinized	3077	2.86	6.08	6.36	6.57	6.92	7.00
		178	2.86	5.89	6.54	6.66	7.05	7.14
	Peridotite serpentinized	1179	2.81	4.81	6.61	6.55	6.82	6.98
		2781	2.74	5.96	6.17	6.67	6.98	7.16
		578	2.88	4.45	7.07	7.03	7.38	7.59
		3177	2.83	5.67	5.87	5.86	6.11	—
		2077	2.71	5.82	6.12	6.21	6.43	6.46
		1277	2.73	5.71	5.84	5.94	6.16	6.30
	Peridotite	977	2.89	5.93	5.52	5.85	6.11	6.22
		2278	2.91	6.29	5.24	5.43	5.71	5.73
	Pyroxenite serpentinized	2783	2.99	6.27	6.61	6.93	7.21	7.34
Baikal	Pyroxenite	379–82	3.05	5.32	6.09	6.20	6.69	6.90
		20–80		5.98	6.16	6.00	6.44	6.37
		439–82	3.03	6.00	5.86	5.76	6.09	—
		380–82	3.21	6.80	7.33	7.41	8.12	—
	Olivine pyroxenite	386–82	3.06	4.31	4.92	4.92	5.36	5.50
		260–81	3.08	5.00	5.04	5.15	5.48	5.59
		34–80	2.89	4.64	4.94	4.82	4.96	4.92
		293–80	3.10	5.34	5.25	5.27	5.33	5.45
		5–83	3.06	5.08	5.04	5.13	5.24	5.36
Kamchatka	Peridotite, pyroxenic	1381	3.190	3.50	5.05	5.82	6.08	6.23
			3.180	3.90	5.20	5.83	6.11	6.30
			3.170	4.70	6.20	7.32	7.59	7.75
			3.190	5.10	6.70	7.59	7.89	8.05
			3.210	4.50	5.70	6.77	7.05	7.20
			3.210	4.50	6.00	6.85	7.17	7.35

conditions (Fig. 1). In particular, the values of longitudinal wave velocities in the pressure range 0.1–1500 MPa for ultrabasites of the Minor Caucasus are in the region limited by the values 5.23–7.07 km/s with the pressure of 100 MPa; 5.42–7.03 km/s with the pressure of 500 MPa; 5.70–7.27 km/s with the pressure of 1000 MPa and 5.72–7.57 km/s with the pressure of 1500 MPa. And the average increase of longitudinal wave velocity in the mentioned pressure ranges is 0.42 m/s MPa.

The average value of the studied rock density, measured under normal conditions, is 2.827 g/cm³. Consolidation at pressure increase was small and we didn't take it into account.

For ultrabasites of Sayany-Baikal folded zone the values of longitudinal wave velocities in the pressure range of 0.1—1500 MPa embrace wider region limited by the values 4.63—7.37 km/s with the pressure of 100 MPa; with the pressure 500 MPa — 4.84—7.75 km/s; with the pressure — 1000 MPa — 4.95—8.15 km/s and with the pressure of 1500 MPa — 5.02—8.48 km/s.

The density of Baikal ultrabasites is higher than in the group of the previous rocks and is 2.99 g/cm³ on the average.

The average increase of longitudinal wave velocities at samples loading in the mentioned pressure range is also higher and makes up 0.5 m/s MPa.

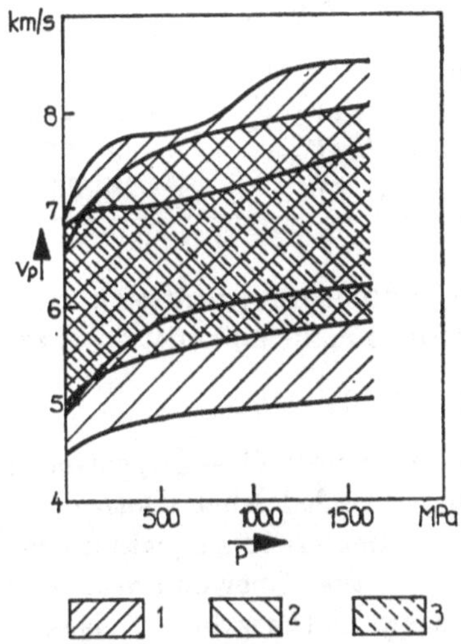

Fig. 1. Regions of distribution of longitudinal wave velocity values for ultrabasites from particular geologic regions.
1 — Baikal; *2* — Kamchatka; *3* — Lesser Caucasus

The ultrabasite samples in Koriak-Kamchatka folded region by the longitudinal velocity values [18] occupy the region limited by the values 4.95—6.80 km/s with pressure 100 MPa; 5.75—7.52 km/s with pressure 500 MPa; 6.00—7.90 km/s with pressure 1000 MPa and 6.20—8.00 km/s with pressure 1500 MPa. The average increase of velocity for these samples in the studied range of pressures is 0.8 m/s MPa.

The rocks of this group proved to be the most dense. The average density is 3.19 g/cm³.

Having compared the obtained values of the physical characteristics in ultrabasites the investigated regions we can make a conclusion that the samples of the Minor Caucasus, Sayany-Baikal and Koriak-Kamchatka folded regions are similar in the values of physical parameters. Only ultrabasites of the Minor Caucasus are less dense than harzburgites of Sayany-Baikal and Koriak-Kamchatka folded regions. Deconsolidation of ultrabasites in the Minor Caucasus is explained by the effect of these rock serpentinization. To take into account this effect serpentinized samples of ultrabasites have been studied (Table 2). The results of these investigations have been processed to reveal the correlation of density value and serpentinization degree [19]. The correlative analysis of the obtained data has revealed that there is a rather steady correlation between

Table 2. Temperature effect on the values of elastic wave velocities of ultrabasites at different pressures

Rock type (NN of a sample)	Pressure (MPa)	Parameter	Values of elastic wave velocities (km/s) at different temperatures (°C)						
			20	50	100	150	200	250	300 °C
Peridotite, serpentinized (879)	300	V_p	6.17	6.09	6.02	5.99	5.95	5.90	5.83
		V_s	3.44	3.42	3.39	3.37	3.32	3.27	3.22
	500	V_p	6.22	6.18	6.16	6.14	6.13	6.12	6.10
		V_s	3.48	3.49	3.45	3.41	3.38	3.37	3.36
Dunite, serpentinized (579)	300	V_p	6.86	6.80	6.75	6.66	6.57	6.35	6.18
		V_s	3.25	3.23	3.22	3.20	3.17	3.11	3.05
	500	V_p	6.89	6.88	6.87	6.86	6.85	6.80	6.75
		V_s	3.28	3.28	3.27	3.24	3.20	3.16	3.14

ultrabasite density and the content of serpentine in them with correlation coefficient $r = 0.80 \div 0.99$, what allowed to make an equation of regression:

$$\varrho = -0.004 \cdot C + 2.95, \tag{1}$$

where ϱ — ultrabasites density (g/cm³), C — serpentine content (%).

The obtained equation has revealed what density the ultrabasites had before the serpentinization process. In particular, this equation has shown that the average density of unchanged ultrabasites of the Minor Caucasus is 3.01 g/cm³.

The effect of high temperature on the values of elastic characteristics of ultrabasites is determined by thermodynamic effect on the samples from the Minor Caucasus ophiolitic association. The experiments have been carried out in two regimes: the first

Table 3. Density of serpentinized ultrabasites

Rock type	NN of samples	Degree of serpentinization (%)	Density initial (g/cm³)	Density with correction (g/cm³)
Dunite serpentinized	178	27	2.864	2.979
	3077	35	2.861	3.002
Peridotite	2278	12	2.905	2.966
	977	26	2.885	2.989
Peridotite serpentinized	3177	49	2.825	3.070
	2077	49	2.724	2.959
	1277	52	2.730	2.990
Serpentinite	1378	99	2.579	3.074
	1178	98	2.462	2.952
	375	85	2.694	3.119
	2477	90	2.622	3.072

regime — under the conditions of samples gradual heating up to t^0 — 350 °C with constant high pressure of 500 MPa; the second regimesamples cooling to normal temperature with the pressure of 300 MPa. The results of the experiments have shown that in ultrabasite rocks the high temperature decreases the values of longitudinal wave velocity (Fig. 2). The gradient of decrease is 1.80 $\div$ −0.45 m/s grad, i.e. taking into account the depth of the earth's crust section gradient of increase of longitudinal wave velocity value from the increase of pressure with the depth in ultrabasites is nearly completely compensated by gradient of velocity decrease by temperature increase. Thus, the values of physical parameters of ultrabasites samples, which have not undergone the processes of weathering and secondary transformation, can be considered as real, expected in natural occurrence at proper depths of the earth's crust and upper mantle. In our case taking into account the fact that in real rocks in the pressure range — 0.1—500 MPa dilatancy microfractures and weathering joints are closed and "healed" it is advisable to use the values of physical characteristics obtained with pressures — from 500 to 900 MPa.

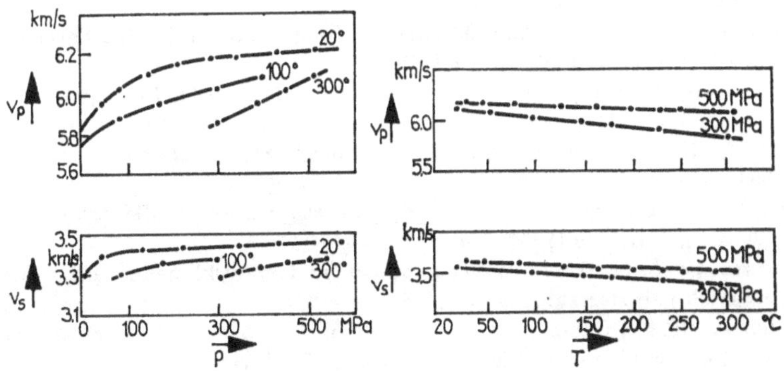

Fig. 2. Isotherms and isobars of elastic waves for a sample N 879 of serpentinized peridotite

In conclusion it can be said that the comparative analysis of the physical characteristics study results of ultrabasites in three geologic regions remote from each other has revealed that the values of physical parameters of rocks samples are very close to each other. This may testify in favour of supposition that the conditions and the mechanism of these rocks formation are analogous. It's necessary to carry out further geologic and geophysical investigations to clear up these conditions.

References

[1] Taylor, H. P.; Noble, J. A.: Sopostavlenie ultraosnovnykh kompleksov Yugo-vostotchnoi Alyaski i drugich chastei Severnoi Ameriki i vsego mira v tzelom. MGK, XXI sessiya, Dokl. sov. geologov, "IL" (1963).

[2] Sheinman, Yu. M.: Magmy i geosynclinalnyi process. Tipy zemnoi kory i magmy. In: Tektonika, magmatism i zakonomernosti razmechsheniya rudnych mestorozhdenii. Nauka (1964).

[3] Shikhalibeili, E. Sh.; Allakhverdiev, G. I.: Ofiolitovye progiby Malogo Kavkaza i ich formirovanie. In: Evolyutziya ofiolitovych compleksov. UNTs, AN SSSR (1981).

[4] SOBOLEV, N. D.: Ultraosnovnye massivy i mantiya Zemli. In: Petrographicheskiye formatzii i problemy petrogenezisa. Nauka (1964).

[5] MORKOVKINA, V. F.; ARUTYUNIAN, G. S.; GAVRILOVA, S. I.: Nekotorye voprosy petrogenezisa ofiolitovoy associatzii Malogo Kavkaza. In: Magmaticheskiye formatzii Kavkaza, yuga Vostochno-Evropeiskoi platformy. Nauka (1977).

[6] KNIPPER, A. L.: Istoriya razvitiya serpentinitovogo melanzha Malogo Kavkaza. Geotectonika. (1971) 6.

[7] PEIVE, A. V.: Okeanicheskaya kora geologicheskogo proshlogo. Geotectonika (1969) 4.

[8] SOKOLOV, S. D.: Olistostromovye tolchshi i ofiolitovye pokrovy Malogo Kavkaza. Nauka (1977).

[9] LUTZ, B. G.: Sostav tektonosfery i ego evolutziya. Nauka (1975).

[10] GENSHAFT, YU. S.: Experimentalnye issledovaniya v oblasti glubinnoi mineralogiya i petrologii. Nauka (1977).

[11] MORKOVKINA, V. F.: Petrologiya giperbasitov Sevanskogo khrebta. Nauka (1974).

[12] PEIVE, A. V.; BOGDANOV, N. A.; KNIPPER, A. L.; PERFILIEV, A. S.: Ofiolity: sovremennoe sostoyanie i zadachi issledovaniya. Geotektonika (1977) 6.

[13] KNIPPER, A. L.: Okeanicheskaya kora v structure Alpiiskoi skladchatoi oblasti (Yug Evropy, zapadnaya chast Azii i Kuba), Nauka (1975).

[14] AUMENTO, F.; LOUBAT, H.: The Mid-Atlantic Ridge near 45° N, 16: Serpentinized ultramafic intrusions. Canad. J. Earth Sci. 8 (1972) 6.

[15] CANN, J. R.: Geological processes of mid-oceanic ridge crests. Geophys. J. Roy, Astron. Soc. **15** (1968) 331–341.

[16] HESS, H. H.: History of ocean basins. In: Petrologic Studies. Buddington Vol., New York (1962) 599–620.

[17] MELSON, W. G.: Petrologic model of the Earth's crust across the Mid-Atlantic Ridge. Trans. Amer. Geophys. Union **49** (1968) 364.

[18] Spravochnik po fizicheskim svoistvam mineralov i gornykh porod pri vysokikh termodinamicheskikh parametrakh. Nedra (1978).

[19] BALAKISHIEV, SH. A.; GASANOV, A. B.; ISAEVA, M. I.: Otzenka vliyaniya vtorichnykh izmenenii na fizicheskie kharacteristiki pri vysokikh p,T-usloviyakh (in press).

Indirect indications change of rock deformation at thermodynamic loading

A. B. GASANOV

In geophysical research practice of objects and phenomena, inaccessible to direct study, their indirect indications and characteristics allowing to obtain with sufficent degree of proved necessary research information are frequently used. For studies of deformation processes of rocks at external loadings and occuring by that redistribution of internal charges in sample volume, such indirect indications are elastic characteristics — longitudinal and transversal waves velocities, ratio of these velocities, modulus of volume compression, shear modulus, Young's modulus and Poisson's ratio. These parameters reflect the interaction character of micro- and macrostructural rock components quantitatively and qualitatively and on the whole — the substance state in deformation process.

At this, bulk and shear modulus characterize phenomena resulting from complicated internal tension in the process of all-round compression of rock, leading to their form and volume changes.

The Young's modulus characterizes the value of the sample stretch depending on their direct resistance to the effect of external loading.

Poisson's ratio, describing the lateral deformation, carries out an information about the availability and character of internal tension display of the samples, and qualifies the intensity of the development of processes of elasticity and plasticity in deformed objects.

Velocity ratio changes of longitudinal and transversal waves may serve as qualitative type indication and macrodeformation development in rocks.

In accordance with the above said, investigations were made in the complex change of rock elastic parameters at different thermodynamic samples. For this purpose experiments were carried out with the rocks of different type on high quasihydrostatic heads device with simultaneous external heating. In the process of experiments there have been registered the longitudinal and transversal wave velocities, and also the linear deformation of samples. The rest elastic modula and coefficients were calculated on the formulas of elasticity theory describing interaction between the components of tension tensors and final deformations [1, 2].

For estimation of values of elastic parameters obtained by the given method, there have been carried out serial experiments with axial deformation of rock samples, allowing to determine the elastic modula by traditional static method. Besides there were carried out rating determinations of modula and hydrostatic compression.

Table 1. Elastic characteristic of rocks under quasi- and hydrostatic pressure

Type of rock	NN of sample	Parameter	Values of elastic characteristics V_p, V_s, E, σ, μ, K, V_p/V_s at hydrostatic/quasihydrostatic pressures (MPa)				
			100	200	300	400	500 MPa
Gabbro-Norite	679	V_p	6.96/6.81	7.12/6.93	7.22/7.11	7.29/7.28	7.35/7.42 km/s
		V_s	3.64/3.61	3.70/3.69	3.73/3.73	3.76/3.74	3.79/3.77 km/s
		E	1.07/1.05	1.11/1,09	1.13/1.12	1.15/1.14	1.17/1.16 GPa 10^2
		σ	0.31/0.30	0.32/0.30	0.32/0.31	0.32/0.31	0.32/0.31 –
		μ	0.40/0.40	0.41/0.40	0.42/0.41	0.43/0.42	0.44/0.43 GPa 10^2
		K	0.95/0.90	1.00/0.93	1.04/0.98	1.05/1.01	1.07/1.02 GPa 10^2
		V_p/V_s	1.91/1.89	1.92/1.89	1.93/1.88	1.94/1.89	1.94/1.92 –
Pyroxene-Gabbro	1579	V_p	7.11/7.06	7.27/7.21	7.37/7.34	7.46/7.47	7.54/7.59 km/s
		V_s	3.87/3.93	4.03/4.02	4.16/4.11	4.23/4.18	4.26/4.23 km/s
		E	1.52/1.54	1.62/1.62	1.72/1.68	1.76/1.74	1.77/1.78 GPa 10^2
		σ	0.28/0.29	0.28/0.28	0.28/0.27	0.28/0.27	0.28/0.27 –
		μ	0.59/0.60	0.63/0.63	0.67/0.65	0.69/0.67	0.70/0.70 GPa 10^2
		K	1.18/1.14	1.22/1.19	1.24/1.22	1.25/1.26	1.27/1.30 GPa 10^2
		V_p/V_s	1.82/1.80	1.80/1.79	1.78/1.79	1.78/1.79	1.78/1.79 –
Serpentinized Dunite	579	V_p	6.54/6.77	6.65/6.84	6.73/6.86	6.80/6.88	6.89/6.89 km/s
		V_s	3.27/3.21	3.29/3.23	3.30/3.25	3.31/3.26	3.33/3.28 km/s
		E	0.78/0.77	0.79/0.77	0.80/0.78	0.81/0.79	0.82/0.80 GPa 10^2
		σ	0.33/0.35	0.34/0.35	0.34/0.35	0.34/0.35	0.35/0.36 –
		μ	0.29/0.28	0.29/0.29	0.30/0.29	0.30/0.29	0.30/0.29 GPa 10^2
		K	0.78/0.88	0.82/0.90	0.85/0.90	0.87/0.91	0.90/0.91 GPa 10^2
		V_p/V_s	1.99/2.08	2.00/2.11	2.00/2.11	2.00/2.11	2.00/2.10 –
Serpentinite	1379	V_p	4.32/4.36	4.36/4.40	3.48/4.41	4.41/4.41	4.44/4.41 km/s
		V_s	2.58/2.55	2.59/2.57	2.61/2.58	2.62/2.58	2.64/2.59 km/s
		E	0.41/0.40	0.41/0.41	0.41/0.41	0.42/0.41	0.42/0.41 GPa 10^2
		σ	0.22/0.24	0.22/0.24	0.23/0.24	0.23/0.24	0.23/0.23 –
		μ	0.16/0.16	0.17/0.16	0.17/0.16	0.17/0.16	0.17/0.17 GPa 10^2
		K	0.25/0.26	0.25/0.26	0.25/0.26	0.25/0.26	0.25/0.26 GPa 10^2
		V_p/V_s	1.67/1.71	1.68/1.71	1.68/1.71	1.68/1.70	1.68/1.70 –

The dependences of enumarated elastic parameters from the value of thermodynamic effect for some investigated samples are given in Tables 1, 2 and Fig. 1, 2. As follows from observation of these data, the character of the change of velocity values of longitudinal and transversal waves with pressure increase has no characteristic anomalies and consists, as a rule, only of two zones – initial with considerable gradient change, and the following gentle zone with low gradient. At increasing temperature the sequence of these zones changes to the opposite, i.e. higher gradient changes of velocities are characteristic for higher temperatures. For some samples (serpentinized dunit and serpentinite) the temperature effect almost compensates the pressure effect.

Explanation to observe regularity of changes of elastic waves is known [3] and is generally related to the remocal of cracks existing in the samples at the beginning of loading, and also arising in them under the influence of pressure of the system. Such

Table 2. Elastic characteristics of rocks under high (500 MPa to 5 kbar) pressure and different temperatures

Type of rock	NN of sample	Parameter	Values of elastic characteristics V_p, V_s, E, μ, K, σ, V_p/V_s at 500 MPa pressure and temperatures, °C							
			20	50	100	150	200	250	300	0°C
Gabbro-Norite	679	V_p	7.42	7.37	7.33	7.33	7.26	7.25	7.24	km/s
		V_s	3.77	3.79	3.80	3.81	3.80	3.80	3.79	km/s
		E	1.16	1.16	1.17	1.16	1.16	1.16	1.15	GPa 10^2
		σ	0.43	0.43	0.44	0.44	0.44	0.44	0.44	–
		μ	0.31	0.31	0.31	0.31	0.31	0.31	0.31	GPa 10^2
		K	1.02	1.03	1.03	1.03	1.03	1.02	1.02	GPa 10^2
		V_p/V_s	1.92	1.91	1.91	1.91	1.91	1.91	1.91	–
Pyroxene-Gabbro	1579	V_p	7.59	7.55	7.55	7.55	7.55	7.55	7.47	km/s
		V_s	4.23	4.20	4.15	4.10	4.05	4.00	3.95	km/s
		E	1.79	1.77	1.72	1 69	1.66	1.63	1.59	GPa 10^2
		σ	0.27	0.27	0.28	0.29	0.30	0.30	0.31	–
		μ	0.70	0.69	0.69	0.66	0.64	0.62	0.61	GPa 10^2
		K	1.31	1.31	1.32	1.35	1.36	1.37	1.37	GPa 10^2
		V_p/V_s	1.39	1.40	1.42	1.44	1.46	1.48	1.49	–
Serpentinized Dunite	579	V_p	6.89	6.88	6.87	6.86	6.85	6.80	6.75	km/s
		V_s	3.28	3.28	3.27	3.24	3.20	3.16	3.14	km/s
		E	0.80	0.80	0.79	0.78	0.77	0.75	0.75	GPa 10^2
		σ	0.35	0.35	0.35	0.36	0.36	0.36	0.36	–
		μ	0.29	0.29	0.29	0.29	0.28	0.28	0.27	GPa 10^2
		K	0.90	0.90	0.91	0.91	0.91	0.90	0.89	GPa 10^2
		V_p/V_s	2.10	2.10	2.10	2.12	2.14	2.15	2.15	–
Serpentinite	1379	V_p	4.41	4.42	4.41	4.39	4.32	4.18	4.05	km/s
		V_s	2.59	2.60	2.59	2.57	2.54	2.51	2.48	km/s
		E	0.42	0.42	0.41	0.40	0.40	0.38	0.37	GPa 10^2
		σ	0.23	0.23	0.23	0.24	0.24	0.22	0.20	–
		μ	0.17	0.17	0.17	0.16	0.16	0.16	0.15	GPa 10^2
		K	0.26	0.26	0.26	0.26	0.25	0.23	0.21	GPa 10^2
		V_p/V_s	1.70	1.70	1.70	1.70	1.70	1.67	1.64	–

forced removal of structural defects leads to the improvement of an acoustic bulk conductivity and increasing of elastic properties. At further loading of rocks the number of newly formed cracks in sample volume begins to prevail, that leads to decrease of the gradient of the value growth of elastic properties. Such peculiar balance of healed and newly formed cracks explains the observed stages of value changes of elastic waves and at heating.

Ratio of elastic velocities at thermodynamic affect changes on more complicated law. As seen from Fig. 1, here may be observed separate anomaly zones. In some cases the effect of temperature is not being compensated, but intensifies pressure effect. Consequently the value of velocity ratio is more sensitive to thermodynamic affect than each of these parameters separately. Thus, it is explained that the velocity of longitudinal

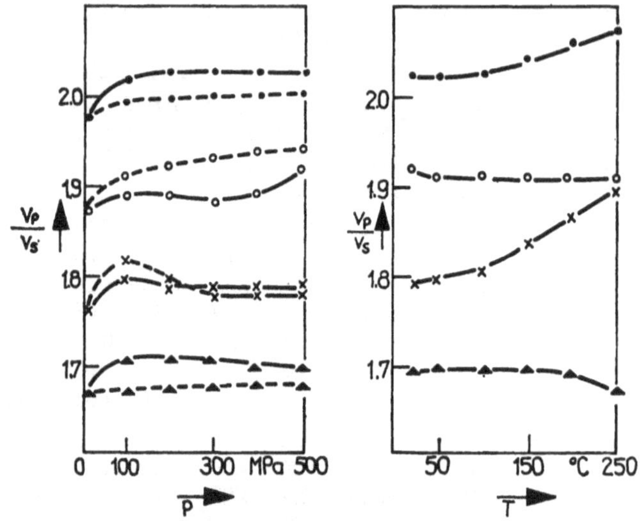

Fig. 1. V_p/V_s of different rocks versus pressure and temperature.
●—● — serpentinized dunite; o—o — gabbro-norite; ×—× — pyroxene-gabbro;
▲—▲ — serpentinite.
(fully line — increasing pressure; dashed line — decreasing pressure)

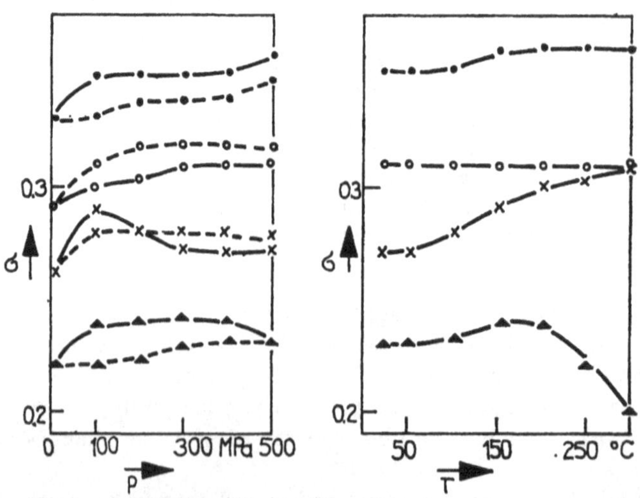

Fig. 2. Poisson's ratio of different rocks versus pressure and temperature.
o—o — gabbro-norite; ●—● — serpentinized dunite; ×—× — pyroxene-gabbro;
▲—▲ — serpentinite.
(fully line — increasing pressure; dashed line — decreasing pressure)

waves depends on medium capacity to resist the strain, but velocity of shear waves — on medium capacity to resist the shearing strain at constant volume [4].

Dependences of Poisson's ratio changes are also characteristic with anomaly zones on the effect of high pressures and temperatures. Taking into consideration that this parameter illustrates plastic, or brittle rock behavior at external loading, and as follows from the given data, has enough sensitivity to thermodynamic affect. Poisson's ratio may be distinguished as the more informational among the indirect deformation indications. It also follows that Poisson's ratio characterizes the capacity of the body to interorthogonal deformations, i.e. on its value change it can be charged about the redistribution of internal tensions in sample volume at high pressures and temperatures. Partially, it follows from our data, that in separate cases joint affect of high pressure and temperature leads to ductility of rocks (Fig. 2). At the same time, for the samples of deliberately ductile rocks such as serpentinites, thermodynamic affect decreases Poisson's ratio, i.e. at high pressures and temperatures the serpentines seek for more brittle state.

Change of the other elastic parameters (shear modulus, Young's modulus, bulk modulus) occurs by analogy to dependences of elastic waves, on linear law with some more gradient of change in the initial zone.

The comparison of elastic characteristics, obtained in quasi- and hydrostatic pressure shows some difference in their values. Aa a rule, difference is observed in the initial zone of pressure, decreasing at pressures higher than 3—4 kbar (Fig. 1, 2). Besides, the observed difference may be of two manner; first — when values of elastic parameters determined in hydrostatic conditions exceed the values of the same, defined at quasihydrostatic conditions; and second — when on the contrary the values corresponding to quasihydrostatic conditions of experiments exceed values corresponding to hydrostatic conditions. By that the first type of difference of elastic characteristics is peculiar to more brittle formations, and second — more ductile. Indeed, more solid and elastic rocks react lesser on predominate axial load in quasihydrostatic pressure chambers, than ductile rocks in which more intensive growth of elastic characteristics in the beginning of loading is observed at quasihydrostatic conditions, when deviation from hydrostatics is more sensitive, i.e. at axial pressure predomination in front of the side. With the pressure growth, heterogeneity of load distribution in working canal of the chamber decreases and approaches to a hydrostatic one, that leads to decrease of value difference of elastic characteristics.

The fact that the change of elastic characteristics of hydrostatic loading have no anomaly zones as at quasihomogenous pressure takes also great notice. It confirms our conclusions that observed anomalies testify the sensitivity of parameters V_p/V_s and σ to processes, connected with the redistribution of internal tensions at quasiheterogenous pressure.

As shown above, for comparative values of investigated parameters there were carried out determinations of elastic modula and by statistical method at uniaxial compression. In particular, there were defined the values of Young's modulus in the zone of linear deformation of samples (Table 3). As was expected, comparison of results of static determination with dynamic ones showed some of their differences, namely — Young's modulus obtained by dynamic method turned out to be higher of static modulus by value. It is explained, as is known [5], by that at dynamic method tensions arising in investigated sample at the affect of elastic supersound impulse are very small, short-timed and practically do not cause irreversible nonelastic processes. While at static experiments of affected tension values essential values are obtained, leading in most cases to definite nonelastic deformations.

Summaring the above said, the following conclusions can be made: From observed elastic parameters (elastic wave velocities, velocity ratios, Young's modulus, shear, bulk and Poisson's ratio) more sensitive and informative occured to be Poisson's ratio and velocity ratio of longitudinal and transversal waves. Consequently, just these characteristics can be observed as indirect indications of rock deformation at thermodynamic loading. Their change of character at external loading depends on elastic properties of rocks and on loading method (hydrostatic or quasihydrostatic). Temperature effect does not always compensate the pressure effect, and in some cases, putting on pressure effect, increases the plasticity or elasticity of rocks.

Table 3. Elastic properties of Pyroxene Gabbro, defined by static methods

NN of samples	Cross section of samples (cm²)	Sample stretch (10^{-3} cm) under unaxial loading (Mp)						Ultimate strength σ_n (10^2 MPa)	E-Modul (static) E_{st} (10^2 GPa)
		0.25	0.5	0.75	1.0	1.5	2.0		
1579_1	1.963	7.6	12.3	16.7	18.5	23.5	25.0	1146	0.555
1579_2	1.980	5.75	10.0	11.3	12.5	13.5	—	1086	0.851
1579_3	1.984	11.5	13.0	14.3	15.4	17.0	20.5	1084	0.553

References

[1] LEBEDEV, T. S. et al.: Uprugiye svoistva gornykh porod pri vysokikh davleniyakh. Kiev, "Naukovo Dumka" (1972).

[2] LEBEDEV, T. S. et al.: Metodika i apparatura odnovremennykh issledovaniy skorosti rasprostraneniya poperechnykh i prodolnykh voln v gornykh porodakh pri vysokikh davleniyakh. Geophys. sb. AN USSR 58 (1974) 41—49.

[3] VOLAROVICH, M. P. et al.: Physico-mekhanicheskiye svoistva gornykh porod i mineralov pri vysokikh davleniyakh i temperaturakh. Moskva, Nauka (1974).

[4] Physicheskiye svoistva gornykh porod i poleznykh iskopayemykh. Moskva, Nedra (1976).

[5] Spravochnik po phyzicheskim svoistvam mineralov i gornykh porod pri vysokikh termodinamicheskikh parametrakh. Moskva, Nedra (1978).

CHAPTER 2

Electrical Properties

Role of water saturation in changes of electric properties of rocks at different *P-T* conditions

T: S. Lebedev, S. I. Shepel

The studies of the effect of water saturation of eruptive and metamorphic rocks in changes of their electric properties at high pressure and temperature are necessitated by water accumulation in upper crust rocks (at least within the first tens of km) which is confirmed by deep and ultradeep wells revealing water-saturated rocks over the whole section as well as electro-magnetic data showing electric resistance values for deep rocks available for laboratory samples only when the latter are saturated with highly mineralized fluids.

At present, only a few works are treating this problem. The laboratory data conclude power dependence of electric conductivity (σ) of crystalline rocks on moisture (ω). When its value does not exceed 0.4% $\lg \sigma = f(\lg \omega)$ curve bends due to significant free-to-bound water ratio variations are observed. The most intensive electric conductivity elevation with increasing NaCl concentration is found for solutions with low salt content. The water saturation degree also influences the character of crystalline rock σ changes with pressure increase, whereas the $\sigma = f(P)$ curves [1] are unaffected by NaCl concentration changes in solutions. For such cases, studies of simultaneous T (to 200 °C) and P (to 2.0 GPa) effect on electric conductivity of water-saturated samples did not reveal any significant ϱ changes with depth [2].

In order to adapt the laboratory experimental conditions closer to those of the interior, the Earth's substance physical properties Department, Inst. Geoph. Ac. Sci. Ukr. SSR (Kiev) performs experimental *P-T* investigations in electric properties of rocks on water-saturated samples using special high pressure chambers enabling us to avoid sample dehydration to 300 °C and 0.7 GPa.

Rocks contain free and bound water. The former is retained in pores and microscopic cracks by capillary forces. Also contained is physically and chemically bound water. The former is retained at solid mineral substance surfaces by molecular coherence forces. The latter is contained by minerals as crystallized (H_2O molecule) and constitutional (hydroxyl group OH^{-1}) water.

By absolutely dry or simply dry samples we will understand rock samples dried at 105–110 °C. Air-dry rocks contain hygroscopic moisture as vapoury water adsorbed from air under atmospheric conditions. Water-saturated samples contain maximum moisture reached by their vacuum-saturation during several days and nights. Water-saturation of rocks under conditions of their natural occurrence is their natural moisture.

Our experiments were mainly made on samples of granitoids (grey fine- to medium-

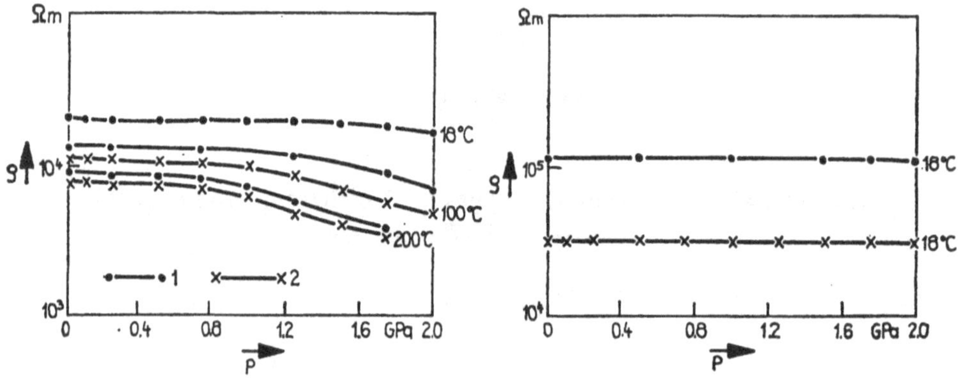

Fig. 1. P-variations of ϱ of fluid-saturated granitoids. a) porphyroblastic migmatite, b) porphyraceous granite. *1* — dc-measurements, *2* — ac-measurements.

grained, pegmatoid, gneissic, porphyroblastic), gneisses and porphyrites of the Ukrainian Shield as well as basaltoids, dolerites and gabbro of the northwester Indian ocean taken by dragging within the Arabia-Indian rift in the area of Vernadsky mountain and on the flange of the Maria Celesta fault [4].

Water-saturation significantly decreases the electric resistance (ϱ) of rocks. Most of ϱ values of dry, air-dry granitoids and those saturated with distilled water fall within 10^8-10^{10}, 10^5-10^7 and 10^4-10^5 Ohm · m respectively. Thus, saturation of granitoids with distilled water decreases electric resistance by 4—5 orders (Table 1). For dry samples a clear differentiation of electric resistance and dielectric constant has been established being less significant when samples contain even a little amount of water (the moisture of air-dry samples did not exceed 0.1%). The differentiation mentioned almost completely disappears for granitoids saturated with distilled water. The effect saturation with distilled and mineralized water has on changes in electric resistance of basalts, dolerite and gabbro also been studied (Table 2). The highest ϱ values are those of air-dry samples. Their saturation with distilled water decreases the resistance more significantly (almost by two orders) for more resistant samples. The least resistant ($2.30 \cdot 10^3$) basalt sample exhibits the least (by ca. 0.1 orders) ϱ decrease when saturated

Table 1. Mean values of electric characteristics of USH rocks with different water content (ϱ = electric resistance (Ωm); ε = dielectric constant, measured at 700 KHz)

Rocks	Number of samples	Dry		Air-dry		Water-saturated
		ϱ	ε	ϱ	ε	
Fine- and medium-grained granites	10	$4.95 \cdot 10^{10}$	5.76	$2.07 \cdot 10^7$	6.95	—
Gneisses	25	$6.87 \cdot 10^9$	6.21	$1.61 \cdot 10^7$	8.19	$5.91 \cdot 10^4$
Gneissic granites	15	$6.18 \cdot 10^9$	6.91	$7.94 \cdot 10^6$	8.44	$1.74 \cdot 10^4$
Pegmatoid granites	14	$3.44 \cdot 10^9$	7.03	$1.15 \cdot 10^7$	7.98	—
Porphyroblastic migmatites	75	$1.52 \cdot 10^9$	8.19	$7.91 \cdot 10^6$	10.03	$6.54 \cdot 10^4$
Picritic porphyrites	11	$1.51 \cdot 10^7$	17.67	$9.28 \cdot 10^5$	22.47	—

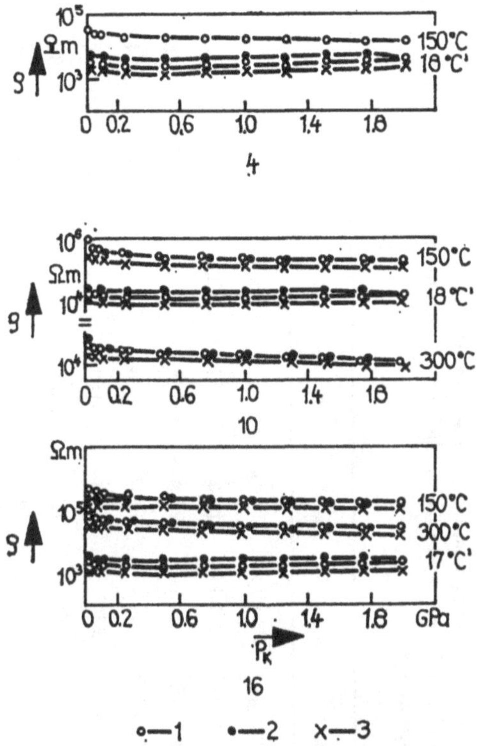

Fig. 2. *P*-dependences of electric resistance of basalts (samples saturated with distilled water — T°C′, dry samples — T°C). *1* — P_q increase, *2* — P_q decrease at dc, *3* — ϱ values at ac.

with distilled water. The lowest electric resistance values are found for basalts saturated with mineralized water. Their ϱ decrease with sea-water saturation (compared to that made with distilled water) may be explained by the occurrence of additional lightly moving ions of salts contained in mineralized water and participating actively in electric conductivity process. The ϱ decrease found is even more pronounced for more (1.66 and $2.09 \cdot 10^6$) than for less resistant basalts for which it is only a few tenths. Of all our dry samples highly and weakly resistance ones are clearly distinguished. This distinct character is also observed for air-dry samples (differing by ca. 2 orders). When saturated with distilled water these basalts diminish the afore mentioned difference to 1 order the latter being almost completely eliminated for basalts saturated with mineralized water. Thus, for the completely saturated rock the crucial effect on ϱ changes is exerted by general amount of fluids in it, number and composition of its current-conducting ions. The role of rock texture and structure of the rock in changes of its electric resistance becomes almost negligible against the prevailing effect of water content. The afore mentioned rules appeared to be valid for dc and ac (1000 c/s) experiments, the ϱ value being however, somewhat lower for ac than for dc in all cases. The greatest difference is observed for air-dry highly resistant rocks (ca. 0.5 orders). For all other samples with electric resistance $< 10^5$ Ohm · m this difference is quite small since electric conductivity is formed by the ohmic current alone at dc but is the sum

Table 2. Electric properties of bottom rocks of Arabia-Indian rift

ϱ^1, ε^1, ϱ^2, ε^2 and ϱ^3, ε^3 are electric resistance ($\Omega \cdot m$) and dielectric constants of air-dry, distilled- and sea-water saturated samples respectively, $\varrho_=$ is electric resistance measured at dc, ϱ_∞ is that measured at ac and 1 kHz, $\varepsilon_{f=1}$ and $\varepsilon_{f=7}$ are dielectric constants measured at 1 kHz and 7 kHz respectively

Rocks	$\varrho^1_=$	ϱ^1_∞	$\varrho^2_=$	ϱ^2_∞	$\varrho^3_=$	ϱ^3_∞	$\varepsilon^1_{f=1}$	$\varepsilon^1_{f=7}$	$\varepsilon^2_{f=1}$	$\varepsilon^2_{f=7}$	$\varepsilon^3_{f=1}$	$\varepsilon^3_{f=7}$
Basalt	$2.09 \cdot 10^6$	$5.91 \cdot 10^5$	$1.68 \cdot 10^4$	$1.42 \cdot 10^4$	$9.43 \cdot 10^2$	$7.54 \cdot 10^2$	27.8	9.98	161	25.1	316	33.1
Basalt	$1.66 \cdot 10^6$	$6.27 \cdot 10^5$	$4.92 \cdot 10^4$	$4.13 \cdot 10^4$	$2.96 \cdot 10^3$	$2.22 \cdot 10^3$	23.9	10.5	69.5	13.4	215	18.1
Basalt	$3.39 \cdot 10^4$	$2.12 \cdot 10^4$	$4.24 \cdot 10^3$	$3.32 \cdot 10^3$	$1.65 \cdot 10^3$	$1.25 \cdot 10^3$	245	23.6	268	23.8	1472	48.0
Basalt	$9.05 \cdot 10^4$	$5.08 \cdot 10^4$	$4.92 \cdot 10^3$	$3.78 \cdot 10^3$	$4.29 \cdot 10^3$	$1.22 \cdot 10^3$	115	11.5	428	29.9	476	34.0
Basalt	$2.30 \cdot 10^3$	$1.93 \cdot 10^3$	$1.38 \cdot 10^3$	$1.14 \cdot 10^3$	$1.15 \cdot 10^3$	$9.17 \cdot 10^2$	2120	34.3	2162	43.5	2360	—
Dolerite	$3.04 \cdot 10^4$	$2.20 \cdot 10^4$	$6.03 \cdot 10^3$	$5.23 \cdot 10^3$	$2.84 \cdot 10^3$	$2.08 \cdot 10^3$	244	18.9	1062	31.9	1102	32.4
Gabbro	$4.37 \cdot 10^5$	$2.25 \cdot 10^5$	—	—	$2.65 \cdot 10^4$	$1.87 \cdot 10^4$	53.5	12.1	—	—	240	16.1

of ohmic and the active component of absorption current at ac. The electric resistance increase at dc compared to that at ac corresponds with the active component value. The latter increases with ϱ elevation, therefore the greatest difference is observed for highly resistant rocks.

Also studied was the water saturation effect on the dielectric constant of these rocks. From Table 2 it follows that the lowest ε values (at different water saturation) are those of the afore mentioned highly resistant basalts and gabbro. Highly resistant basalts and gabbro saturated with sea water show dielectric constants of 18.1 to 33.1 and 16.1 (at 7 kc/s) respectively. Under the same conditions less resistant basalts (3.39 and $9.05 \cdot 10^4$ Ohm · m) exhibit ε values of 48.0 and 34.0 respectively. At 1 kc/s (1 kHz) these differences in the dielectric constant are even more marked. Thus, the most interesting result of these experiments is that, unlike electric resistance, ε is a structure-sensitive parameter both for dry samples and for those saturated with highly mineralized fluids.

Quasi-hydrostatic pressure elevation to 2.0 GPa decreases electric resistance of water-saturated samples of granitoids by analogy with dry samples. The ϱ changes mentioned are observed at temperatures 18, 100 and 200 °C. Thus, water does not affect markedly the P-dependence of electric resistance of granitoids. As known, water generally favours ϱ increases with elevating pressure due to greatly serpentinous current-conducting channels, decrease of their cross-section area and decreasing saltion mobility in solution. These effects are of minor importance for water-saturated granitoid samples probably due to low fluid-content of low-porosity. rocks. The main factor decreasing ϱ of water-bearing granites in quasi-hydrostatics is decrease of electric resistance of the rigid mineral skeleton of the rock with increasing pressure. A principal difference in the character of $\varrho = f(P_q)$ dependence for dry and water-saturated basalt samples has been established. Dry samples decrease ϱ to 2.0 GPa, this decrease being to 0.5—0.75 GPa for water-saturated ones increasing it again (at dc and ac) under higher P.

The principal differences found for electric resistance dependence of water-saturated and dry basaltoid samples may probably be explained as follows. In quasi-hydrostatics, the low-pressure non-hydrostatic component is known to affect ϱ decrease most significantly, its effect on ϱ notably decreasing at $P > 0.5$—0.7 GPa. Here rocks of comparatively high porosity are observed to increase their electric resistance by water saturation.

Temperature dependence of electric resistance of water-saturated granitoid and basaltoid samples are characterized by a very low activation (E_0) of the current carriers. For granites, E_0 is 0.12—0.18 eV to 200 °C slightly increasing at higher temperature, which is significantly lower than for dry samples. Water-saturated basalts exhibit even lower E_0 values ranging to ca. 0.1 eV at 100—125 °C, which corresponds with mineralized-water activation energy. The 150—300 °C range shows higher (0.27 to 0.33 eV) E_0 values. However, the activation energy of water-saturated basalt samples within the range of extrinsic conductivity is much lower than that of dry ones (in our case they differ by 0.26 and 0.38 eV).

There has also been studied ϱ of water-saturated granitoids at simultaneous P-T effect using the programmed modelling method [5]. The experimental results enabled us to characterize peculiarities of changes in electric resistance of these rocks with depth. The ϱ decrease by 1.3—1.5 orders for water-saturated granitoid and basaltoid samples was found under laboratory P-T conditions imitating depths to 20 km. Such

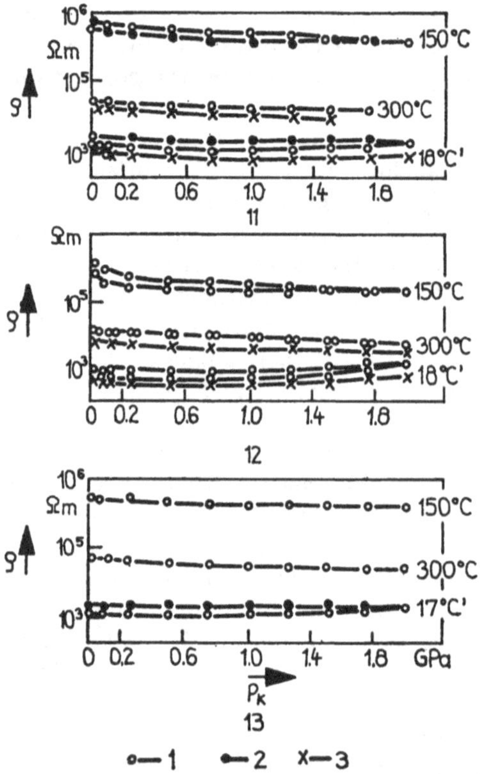

Fig. 3. Quasihydrostatic pressure (P_q) effect on change in electric resistance of basalts saturated with sea water (T°C') and dry ones (T°C) at different temperatures. The legend is the same as in Fig. 2.

a character of ϱ changes with depth is mainly defined by features of electric conductance temperature dependence of rocks.

The character of the ϱ distribution at depths, the ϱ isobars at different pressures and mean values under atmospheric conditions as well as data on compositional changes of the interior enabled us to set up the first versions of normal geoelectric sections of the upper continental (central Ukrainian Shield) and oceanic (Arabia-Indian rift in the Indian ocean) crust. The geoelectric section of the Ukrainian Shield has been set up for depths to 3500 m using Kirovograd deep borehole cross-section data [6]. At a depth of 3 km, pressure and temperature decrease the electric resistance of migmatites, gneisses and granites by 0.3, 0.35 and 0.4 orders respectively, the contribution of pressure to general ϱ decrease being about 0.1 orders. Down to the depth of about 3 km the electric resistance of airdry and water-saturated granite samples is about 10^6 and only 10^4 Ohm · m respectively. This is mainly due to petrographic composition variations in rocks over the borehole section which contribute differently to the ϱ decrease at simultaneous P-T effect. The electric log data of this borehole show its ϱ values to be not more than 10^3 Ohm · m. Their coincidence is only assumed to be possible when rocks are saturated with highly mineralized water under natural conditions, which has

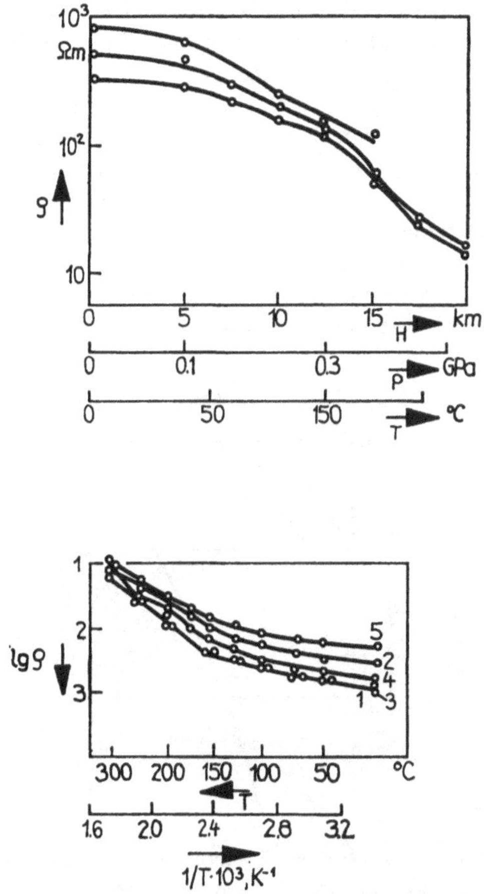

Fig. 4. $\varrho = f(PT) = f(H)$ and $\lg \varrho = f(1/T)$ dependences for samples of basalts saturated with sea water. *1, 2* — 0.0001 GPa; *3* — 0.1 GPa; *4, 5* — 0.5 GPa.

been confirmed by borehole water analyses. However, the electric log data show the ϱ decrease to a few tenths of Ohm · m at a depth of about 2.8 km, which cannot be explained by mineralized water alone. Here, intensive rock fracturing and water invasion may occur.

Also proposed is the first schematic version of a normal geoelectric section of the Arabia-Indian rift area (Fig. 5). As is known, a thin sedimentary layer exists in the oceanic crust. Below it an almost 2 km thick mass of predominantly evolved basaltoids alternating with sediments is clearly seen. At the crust bottom basic rocks (gabbro, metagabbro, dolerites, metadolerites etc) about 5 km thick occurs. Electric resistance of the upper part of the basaltic layer is about 800 Ohm · m notably decreasing under elevating P. The ϱ value of the lower part of the basaltic layer is about 300—400 Ohm · m. Our laboratory data suggest electric resistance of water-saturated gabbro and dolerite samples to be much higher than that of basalts due to low porosity of dolerite and gabbro samples hence their lower sea-water saturation.

At P and T corresponding with depth of the order of 2 km dolerite and gabbro electric resistance decreases to about $5 \cdot 10^3$ Ohm · m. Consequently, transition to basic rocks should elevate ϱ due to compositional and textural changes in bottom rocks. In principle, such a picture corresponds with the $\varrho = f(H)$ curve set up by data given in [7]. Within this layer the deepening decreases the electric resistance due to P and T increase. Also possible is simultaneous gradual rock dehydration causing an ϱ increase, too. Thus, here, under real deep P-T conditions the character of electric resistance changes in oceanic crust mineral matter will be determined by the total effect of these two factors. The former may be rather reliably studied by experimental model thermobaric programs. Such experiments have been made only rarely so far. The problem of changes of oceanic crust fluid regime with depth has been studied insufficiently.

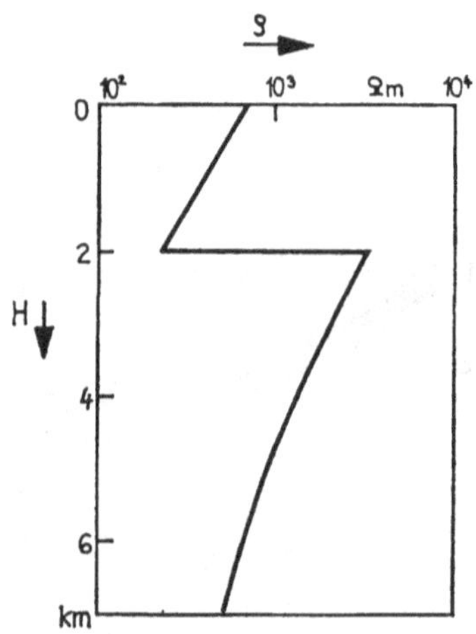

Fig. 5. Schematic normal geoelectric section of Earth's crust of continental type (Arabia-Indian rift and north-western Indian ocean).

Conclusions

1. Water saturation of rock samples under natural P-T conditions decreases their electric resistance from tenths to several orders. The degree of the ϱ decrease of completely dry samples also depends on their initial value.

2. Water accumulated in samples smoothes away the differentiation of rocks based on their electric resistance. Dielectric constant remains a structure-sensitive parameters even at complete rock saturation with water.

3. A similar character of P-dependence of ϱ show dry and water-bearing low-porosity granites. However, it is different for high-porosity basaltoids with high water content. Temperature and thermobaric ϱ changes depend on whether or how much moisture is contained in the samples.

4. On the basis of laboratory experiments with water-saturated rocks the first versions of normal geoelectric sections of the upper parts of the continental and oceanic crust have been set up.

References

[1] Parkhomenko, E. I.; Stefankevich, Z. B.; Vysokova, I. I.: On the role of moisture in electric conductance of crystalline rocks at elevated pressures and temperatures (in Russian). Izv. Ac. Sci. USSR, Fizika Zemli, 5 (1972) 84–91.

[2] Bakiev, Muz. Kh.; Parkhomenko, E. I.; Bakiev, M. Kh.: Investigation of electric conductance of saturated rocks of Tashkent region at pressures and temperatures under conditions of closed system (in Russian). In: Phys. properties of rocks at high thermobaric parameters. Elm, Baku (1978) 180–181.

[3] Lebedev, T. S.; Shepel, S. I.: Thermobaric investigations of electric parameters of granitoids (in Russian). Geof. Zhurn., 6 (1984) 4, 14–23.

[4] Lebedev, T. S.; Shepel, S. I.; Burtny, P. A.; Rusakov, O. M.: Electric characteristics of Indian ocean floor rocks (in Russian). Geofiz. Zhurn., 5 (1983) 1, 45–56.

[5] Lebedev, T. S.: Problems of thermobaric studies of lithosphere's mineral matter physical properties (in Russian). Geofiz. Zhurn., 7 (1985) 6, 62–82.

[6] Lebedev, T. S.; Shepel, S. I.: Electrical properties of granitoids of the Ukrainian Shield under various pressure and temperature conditions. In: The International monograph CAPG: Physical properties of the mineral system of the Earth's interior. Praha, (1985) p. 80–90.

[7] Hyndman, R. D.; Drury, M. J.: The physical properties of oceanic basement rocks from deep drilling on the Mid-Atlantic Ridge. J. Geophys. Res. 81 (1976) 23, 4042–4052.

Petro-electrical properties of granulites and ultrabasites of the saxonian granulite massif

V. Kropáček, M. Laštovičková, M. Kopf

Introduction

The 2nd International Petrophysical Expedition to the Saxonian granulite range was organized as part of CAPG Project 3 (1981). Granulitic and ultrabasic rocks from the Earth's crust — upper mantle boundary can be found in this massif. The electrical conductivity of these rocks was studied under temperatures of up to 1000 °C in an effort to contribute to the interpretation of the electrical conductivity in this important range of depths. The geological situation, rock sampling and geological sketch map of the region can be found by [1].

Method of Measurement

The electrical conductivity of the rock specimens was measured by the double-electrode method in a flowing argon atmosphere ($p_{O_2} \sim 10^{-1}$ Pa) [2]. The temperature of the specimens was measured by a Pt-PtRh thermocouple with an accuracy of ± 2 °C, and the accuracy of measuring the electrical conductivity was about 5% of the measured value. The experiments were conducted with rock specimens of cylindrical shape, diameter 8 mm and height 10 mm.

Experimental Results

In analysing the petro-electrical results, attention was mainly paid to determining and comparing the differences in the values of the electrical conductivity between the separate types of granulites, ultrabasites and also between the granulitic and ultrabasic rocks of this region.

The mean petrographic and chemical composition of the studied rock samples is summarized in Table 1 and the mean values of the electrical conductivity are given in Table 2.

Granulites: The values of the electrical conductivity of the granulites of the Saxonian granulite massif in dependence on temperature are shown in Fig. 1. The statistical t-test, applied to the values of the electrical conductivity at temperatures of 200, 300, 400, 600, 800 and 1000 °C, indicates that the curves of the temperature dependence of electrical conductivity, $\log \sigma(T)$, at the level of significance $p = 0.05$ display statistically

Table 1. Mean petrographic and chemical composition of the studied rocks

Rock type	qz (vol.-%)	af	pl	px	ga	bio	SiO_2 (wgt.-%)	Al_2O_3	Fe_2O_3	MgO	CaO	H_2O
Quartz-Granulite $N = 3$	60	30	—	—	6	1	75	14	3	1	2	—
Granulite $N = 11$	40	43	10	—	4	2	75	13	3	1	1	—
Pyroxenic Granulite $N = 3$	—	—	55	35	—	—	60	12	12	7	9	—
Pyroxenic Granulite (pyriclasite) $N = 2$	—	—	50	45	—	—	50	16	17	7	9	—
Pyroxenite $N = 2$							45	11	15	20	5	2
Ca-rich Pyroxenite $N = 1$	—	—	—	100	—	—	47	12	6	16	20	—
Eclogite $N = 4$	—	—	—	30	60	—	43	14	10	16	11	2
Eclogite (calc-silicate) $N = 1$	—	—	—	70	30	—	46	8	10	10	23	—
Bronzite-Serpentinite $N = 1$							40	1	8	38	1	12
Garnet-Serpentinite $N = 5$							39	4	15	36	2	11
Dunitic-Serpentinite $N = 1$							40	3	15	27	2	8

Table 2. Mean values of the electrical conductivity $\log \sigma$ $[\Omega m]^{-1}$ at various temperatures

Rock type	Number of specimens	1000 °C	800 °C	600 °C	400 °C	300 °C	200 °C
Quartz-Granulite without biotite	1	−3.4	−4.1	−4.6	−7.1	−7.9	−7.5
Quartz-Granulite with biotite	2	−3.3	−3.7	−4.6	−6.7	−8.0	−8.0
Granulite without biotite	4	−3.3	−4.3	−4.9	−6.8	−6.9	−6.9
Granulite with biotite	4	−3.3	−3.6	−4.9	−7.0	−7.7	−7.5
Pyroxenic Granulite $SiO_2 = 60\%$	3	−2.7	−3.0	−4.0	−5.0	−5.5	−5.8
Pyroxenic Granulite $SiO_2 = 50\%$	2	−2.0	−2.7	−3.7	−4.3	−4.5	−4.9
Pyroxenite	2	−2.9	−3.4	−4.2	−5.1	−5.5	−5.9
Ca-rich Pyroxenite	1	−3.3	−4.4	−5.2	−6.8	−7.4	−7.1
Eclogite	3	−1.9	−2.5	−3.8	−5.1	−5.4	−6.0
Eclogite (calc-silicate)	1	−1.3	−1.9	−2.9	−4.5	−4.2	−3.9
Dunitic Serpentinite	1	−1.5	−2.0	−2.3	−3.2	−3.7	−4.2
Garnet-Serpentinite	6	−2.7	−3.4	−4.2	−5.0	−5.6	−6.4
Bronzite-Serpentinite	1	−3.7	−4.4	−5.0	−5.0	−5.2	−5.1

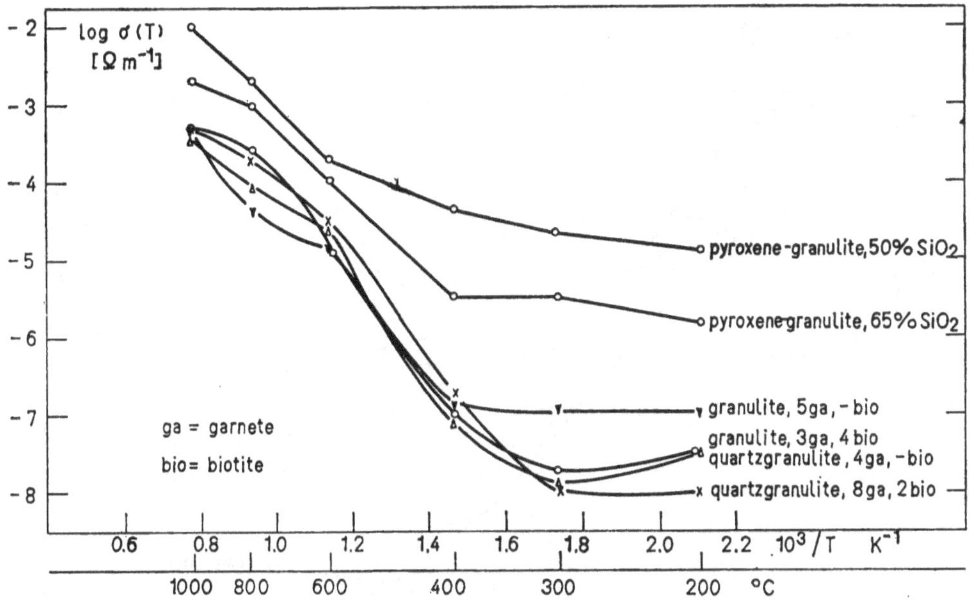

Fig. 1. The electrical conductivity of the granulites of the saxonian granulit massif in dependence on temperature.

significant differences for acid and basic granulites. The experimental results can be characterized as follows:

1) The basic granulites (i.e. pyroxenic granulites, pyriclasites) display statistically significant higher values of electrical conductivity than the acid granulites.

2) Pyroxenic granulites with lower contents of SiO_2, i.e. more basic types, display significantly higher values of electrical conductivity.

3) The following was observed in normal granulites:

- quartz-granulites, containing more quartz than 50 vol.-%, display lower values of electrical conductivity than the granulites containing about 40 vol.-% quartz,
- granulites containing biotite display higher values of electrical conductivity at temperatures of 200–300 °C than analogous granulites which do not contain biotite, however, at 800 °C the values of their electrical conductivity are lower than those of the granulites with no content of biotite,
- at temperatures of 400 ($\log \sigma \sim -7.0$), 600 ($\log \sigma \sim -4.7$) and 1000 °C ($\log \sigma \sim -3.3$) the differences between the temperature curves of electrical conductivity of the specimens of normal granulites are practically negligible.

4) The increase in electrical conductivity with temperature differs greatly for the separate groups of granulites:

- in acid and normal granulites, the increase in electrical conductivity with temperature up to 1000 °C amounts to six orders of magnitude (i.e. from 10^{-8} to 10^{-3}–10^{-2}) $[\Omega m]^{-1}$, in basic granulites this increase is substantially lower and only amounts to three orders (i.e. from 10^{-5} to 10^{-2} $[\Omega m]^{-1}$),

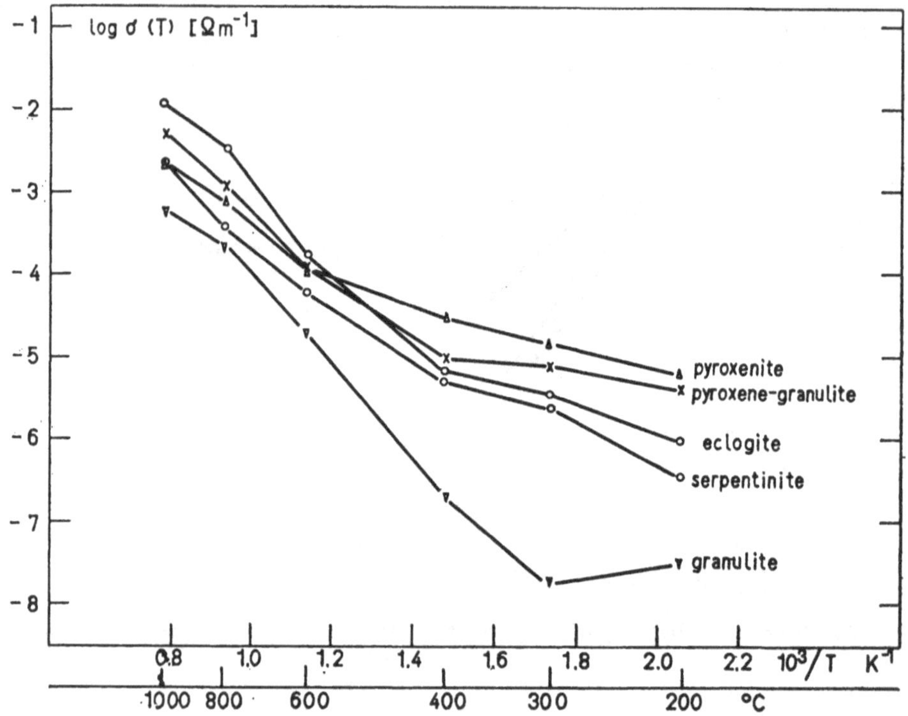

Fig. 2. The electrical conductivity of ultrabasic and basic rocks in dependence on temperature.

- practically no increase in conductivity can be observed in any type of granulite up to 300 °C,
- above 300 °C the electrical conductivity, in terms of $\log \sigma = F(1/T)$, begins to increase practically linearly; this increase is more conspicuous in the acid granulites than in the basic,
- at about 800 °C, it is quite evident that, as compared to 200 °C, the conductivity values of the granulites containing, biotite are lower than those of the granulites with no content of biotite.

5) The values of the activation energy of acid and basic granulites also differ in the temperature range of 400—1000 °C. In this temperature range, the values of the activation energy of pyroxenic granulites and acid granulites are $E = 0.87$ and 1.04 eV, respectively.

Ultrabasic rocks: The differences in the values of the electrical conductivity of the individual ultrabasic and basic rocks in the set being studied are very small, as can be seen in the graphs in Fig. 2. The experimental results obtained for the studied ultrabasic and basic rocks can be summarized as follows:

1) The conductivity values of the basic pyroxenic granulites are within the range of values for pyroxenites, and at higher temperatures also in the range of values for eclogites and serpentinites.

2) In ultrabasic rocks the increase in electrical conductivity with temperature is not

as large as that observed in acid granulites. In the temperature range of 400–1000 °C, the activation energy of ultrabasic rocks is considerably lower, i.e. $E = 0.62$ eV.

3) The increase in electrical conductivity with temperature in ultrabasic rocks is very irregular. Of all the samples studied, serpentinite displays the lowest conductivity values in the whole range of temperatures, i.e. 200–800 °C. It is remarkable that pyroxenite and pyroxenic granulite display higher conductivity values at 200 °C than eclogites, whereas at 600 °C the values of eclogite are already higher than those of the pyroxenic rocks.

Table 3. Electrical conductivity $\log \sigma$ $[\Omega m]^{-1}$ of eclogites of various regions

Region	1000 °C	800 °C	600 °C	400 °C	300 °C	200 °C
Saxony	−1.9	−2.5	−3.8	−5.1	−5.4	−6.0
Bohemian Massif	−2.5	−2.9	−3.8	−5.0	—	—
Siberia[1]	—	−4.3	−5.4	−6.7	−8.3	—
Southern Urals[1]	—	−3.5	−4.4	−5.8	−6.9	—
Kazakhstan [1]	—	−2.8	−3.7	−5.0	−6.0	—

[1] Mean values adopted from Laštovičková and Parchomenko [1978].

4) The mean values of the temperature dependence of electrical conductivity of the Saxonian eclogites overlap with the values of the mantle eclogites from Siberia, of the ophiolitic eclogites from the southern Urals and of the crustal eclogites from Kazakhstan [3]. Also the difference between Bohemian and Saxonian eclogites is insignificant (Table 3).

5) The number of specimens of serpentinite and peridotite was too small to be able to draw more significant conclusions. One can claim that:

- the dunitic serpentinite displays the highest value of electrical conductivity of all the ultrabasic rocks: at 1000 °C its conductivity value is higher than that observed in eclogites,
- after being heated to 600 °C, serpentinites containing about 13 wgt-% H_2O disintegrate,
- the comparison of the electrical conductivity of the serpentinites of the Saxonian granulite massif with the peridotites of the Bohemian Massif indicates that their curves of temperature dependence of electrical conductivity are practically identical between 600 and 1000 °C. The differences observed between 200 and 400 °C can be explained by the lower degree of serpentinization of the peridotites of the Bohemian Massif.

The content of SiO_2 in the rocks and its relation to the values of the electrical conductivity

The relations between the SiO_2 content and electrical conductivity observed at various temperatures are illustrated in Fig. 3. This figure implies the following:

1) In the group of granulitic rocks, the conductivity is observed to increase as the SiO_2 content decreases.

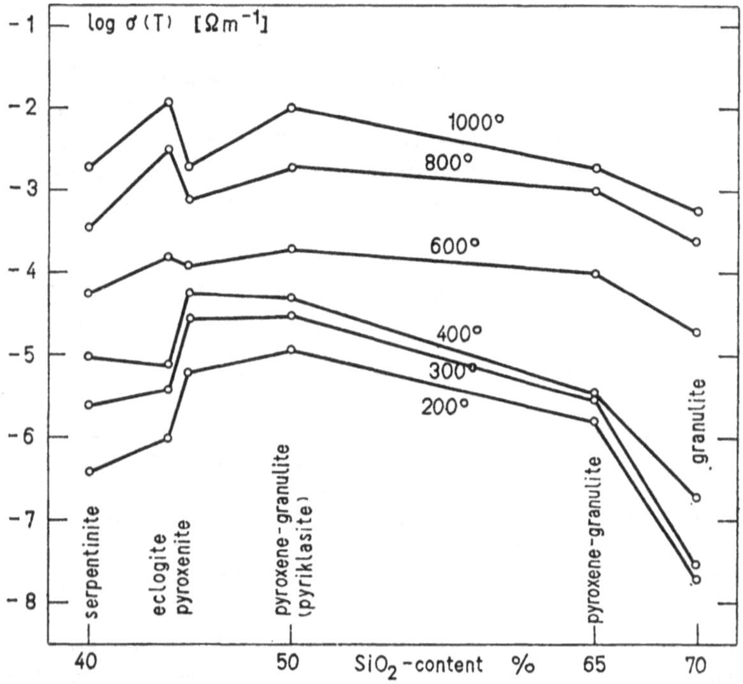

Fig. 3. The electrical conductivity at various temperature in dependence of SiO_2-content.

2) Opposite to the linear relations between the SiO_2 content and other petrophysical parameters (e.g. density, velocity of elastic waves), the electrical conductivity displays a distinctly non-linear relation. The conductivity values at first increase with the SiO_2 content up to about 50 wgt-%. The conductivity then decreases with increasing SiO_2 content and reaches the lowest values for acid granulites with an SiO_2 content of approximately 70 wgt-%. This trend in the conductivity values is observed at temperatures of 200, 300 and 400 °C.

3) At 600 °C, the conductivity values of all the studied rocks (SiO_2 content = 40—70 wgt-%) equalize within one order of magnitude (10^{-4}—10^{-5}).

4) A very interesting increase in conductivity at temperatures over 600 °C is observed in the studied eclogites in dependence on the SiO_2 content. The same increase was also observed in the eclogitic rocks of the Bohemian Massif.

Activation energy

Table 4 gives the values of the activation energy of the studied rock samples in the temperature intervals of 200—400 and 400—1000 °C. In the latter interval, the highest values of activation energy are displayed by acid granulites, the lowest by pyroxenites and pyroxenic granulites.

Table 4. Mean values of the activation energy E_0 [eV] of the studied rock types

Rock type	Temperature interval	
	200–400 °C	400–1000 °C
Quartz-Granulite	0.16	1.04
Granulite	0.21	0.87
Pyroxenite	0.19	0.63
Eclogite	0.21	0.93
Serpentinite	0.41	0.62

Conclusions

Changes in the electrical conductivity of the rocks of the Saxonian granulite massif with temperature were studied in the temperature range of 200–1000 °C. The studied set of rocks contained granulites (acid, basic), pyroxenic granulites, pyriclasites, eclogites, pyrigarnets and serpentinites. At 200 °C, the lowest values of electrical conductivity were displayed by acid granulites ($\sigma \sim 10^{-8}$ $[\Omega m]^{-1}$); serpentinites ($\sim 10^{-6}$), eclogites ($\sim 10^{-6}$), pyriclasites ($\sim 10^{-6}$) and pyroxenic granulites ($\sim 10^{-6}$) have higher conductivity values. At 1000 °C, the differences in the conductivity values are considerably smaller, within the range of 10^{-4}–10^{-2} $[\Omega m]^{-1}$. The relation between the values of the electrical conductivity and the SiO_2 content in the rock at first displays an increase of conductivity with the SiO_2 content, and in rocks containing more than 50 wgt-% SiO_2 the conductivity then decreases as the SiO_2 content continues to increase.

References

[1] Fiala, J.; Vaňková, V.: Radioactivity and geochemistry of granulites and some related rocks from the Saxonian granulite complex. Studia geoph. et geod., **29** (1985) 374.

[2] Laštovičková, M.; Popule, J.: A method of measuring the temperature dependence of the electrical conductivity of rocks. Studia geoph. et geod., **18** (1974) 370.

[3] Laštovičková, M.; Parkhomenko, J. J.: Electrical properties of basalts from the Bohemian Massif under high thermodynamical parameters (in Russian). Fravaux Geophysiques XXIV (1976) 464. Prague, Academia (1978) 289.

Laboratory studies of electrical conductivity of rocks in Czechoslovakia in 1981—85

M. Laštovičková

In the period of 1981—85, the electrical conductivity of sets of rocks (Table 1) and their main rock-forming minerals has been studied in the temperature interval of 200 to 1000 °C (argon atmosphere $p_{O_2} \sim 10^{-1}$ Pa). Choice of rocks, representing the most important building elements of the lower crust and upper mantle, was carried out from the point of view of possible application in geophysical models.

The complex studies treated the connections of electrical properties of rocks with their modal composition, chemistry and their structural parameters with the use of other physical methods, e.g. the measurement of magnetic properties, heat conductivity, differential thermal analysis, thermogravimetry, X-ray analysis and studying the influence of the time-factor on the temperature dependence of electrical conductivity, because only from these measurements we can obtain the real values of electrical conductivity, applicable in modelling. Long-time measurements enable us to explain reversible and irreversible changes in the samples of minerals (rocks) and the dynamics of the whole process of heating.

It follows from Table 1 that the alkaline volcanic rocks of the Bohemian Massif (72 localities) have the highest conductivities. The electrical conductivity of the Slovak-

Table 1. Temperature dependence of electrical conductivity of volcanic rocks (N — number of localities)

Type of rock	Mean log conductivity ($\log \sigma/\Omega m/^{-1}$)			N	Chemically analyzed in
	400 °C	600 °C	800 °C		
Alkaline volcanic rocks of the Bohemian Massif	-3.94 ± 0.90	-3.00 ± 0.63	-2.42 ± 0.51	72	[1]
Teschenites, Trachyande-	-5.26 ± 0.95	-3.22 ± 0.79	-2.49 ± 0.96	41	[4, 5]
sites of Moravia	-6.36 ± 0.59	-4.37 ± 0.61	-3.37 ± 1.08	10	
Alkaline basalts of Poland	-5.15 ± 1.72	-3.75 ± 0.72	-3.22 ± 1.08	13	[6]
Basalts of Bulgaria	-3.97 ± 0.75	-3.22 ± 0.66	-2.58 ± 0.53	12	—
Calc-alkaline, alkaline basaltic rocks of Slovakia	-5.60 ± 1.08	-4.21 ± 0.85	-3.32 ± 0.82	56	[2, 3]

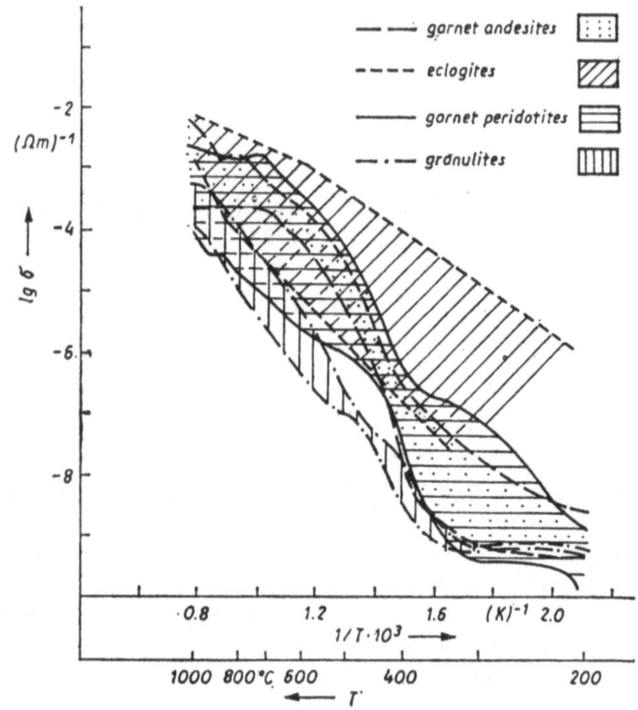

Fig. 1. Temperature dependence of electrical conductivity for garnet rocks.

ian rocks, the second most frequent group of rocks (56 localities), belongs to lower conductivities among the mentioned sets of volcanic rocks over the whole range of temperatures. Admixture elements have a pronounced effect on their conductivities. In our older studies the tendency of increasing the electrical conductivity with the increasing content of $\sum$ Fe from FeO + Fe_2O_3, TiO_2, or Ni, Cr, Co was observed. For these experimentally established tendencies, we tried to find a mathematical description for the increase of the conductivity value with the content of various components. We applied the percolation theory to the set of 51 alkaline volcanic rocks from the Bohemian Massif and 16 calc-alkaline rocks from Slovakia with widely varying contents of TiO_2 (0.1—6.0) wt-% and of $\sum$ Fe (2.5—17.0) wt-%. Our calculation [1] indicates that the percolation theory describes the behaviour of conductivity very well below 500 °C, but not above this temperature.

The electrical conductivity of rocks of the teschenite formation and trachyandesites of Moravia is lower over the whole range of temperatures than that of the volcanic rocks of the Bohemian Massif. As regards chemical composition, trachyandesites represent a link between the Bohemian Massif and the volcanic rocks of the Carpatian System. We observed that the teschenites displayed dissociation of various carbonate minerals (500—700 °C) which was confirmed by measuring the AC conductivity [5].

The electrical conductivity of Fe—Ti—O minerals, the main rock-forming minerals of basaltic rocks, was investigated in [7]. Generally, the electrical conductivity was found to decrease with increasing degree of oxidation.

Fig. 1 shows the range of conductivity values for rocks representing the upper

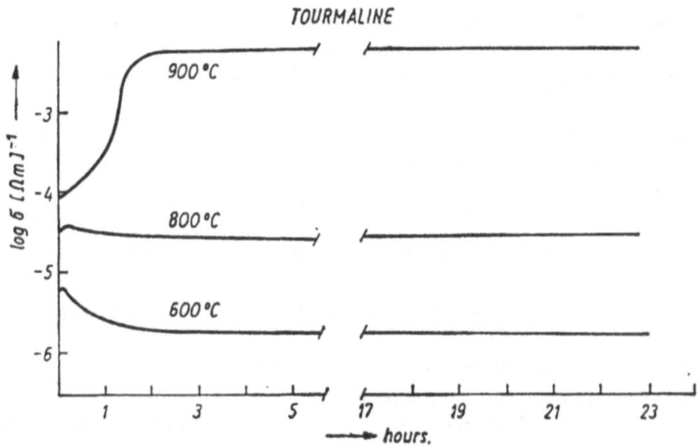

Fig. 2. Time-temperature dependence of electrical conductivity for tourmaline.

mantle — garnet peridotites, eclogites and granulites from the Bohemian Massif and garnet andesites from Central Slovakia [8]. The temperature dependence of electrical conductivity of various types of natural garnets, which are the main rock-forming minerals of these rocks, was investigated also in [8]. The electrical conductivity is strongly dependent on the content of almandine component, therefore the conductivity of almandines was studied also under the influence of time factor [9].

As regards minerals, it follows from our testing methods that it is necessary to use the conductivity data of minerals directly separated from these rocks for analysis. The conductivity of these minerals does in fact differ slightly (not in order of magnitude) from the conductivity of synthetic minerals (as a result of impurities, etc.).

As a base of studying very large expansed group of intrusive rocks of the Earth's crust — group of granites — the complex analysis of their main rock-forming minerals (plagioclase, amphibole, orthoclase, muscovite, tourmaline) was performed. Fig. 2 is an example of possible changes of electrical conductivity of tourmaline under the influence of time-factor. These changes of electrical conductivity were explained by means of changes registrated in differential thermal analysis and thermogravimetry.

The substantial part of results of laboratory measurement of electrical properties of rocks and minerals, performed in 1981—1985 in the Geophysical Institute in Prague and their discussion with connection to world's data are presented in the review paper [11].

References

[1] Laštovičková, M.; Kropáček, V.: The Experimental Screening of the Percolation Theory on the Alkaline Volcanic Rocks of the Bohemian Massif, Geof. Zhurnal Akad. Nauk USSR 5 (1983) 77—81 (in Russian).

[2] Laštovičková, M.; Kropáček, V.: Electrical Conductivity of Young Basaltic Rocks from Central and SE Slovakia. Studia geoph. et geod. 24 (1980) 389—399.

[3] Laštovičková, M.: Electric Properties of Neovolcanic Rocks of Central and East Slovakia. Travaux Géophysiques, XXIX, (1981) 558, 217—233.

[4] Laštovičková, M.: Electrical Conductivity of Teschenites, Trachyandesites and Basalts under High Temperatures. Travaux Géophysiques XXVII (1982) 529, 233—250.

[5] Laštovičková, M.; Avaliani, Z. S.; Geladze, G. G.: Electrical properties of teschenites from Czechoslovakia under high temperatures (in Russian). Physical Properties of Rocks at High Pressure and Temperature for Seismological Application. Tashkent, Fan (1981) 125 to 126.

[6] Jelenska, M.; Kadzialko-Hofmokl, et al.: Magnetic Properties of Basaltic Rocks of Lower Silesia and their Relationship to Petrological Characteristics. Čx. čas. pro mineralogii a geologii **23** (1978) 159—180.

[7] Kropáček, V.; Laštovičková, M.: Changes of Structure, Phase Composition and Electric Conductivity under High-Temperature Oxidation of Titanomagnetites. Journ. Geophys. **48** (1980) 40—46.

[8] Laštovičková, M.: Electrical Conductivity of Garnets and Garnet Containing Rocks. Gerlands Beitr. Geophys. **48** (1981) 529—537.

[9] Laštovičková, M.: Temperature — Time Dependence of the Electrical Conductivity of Garnets. Studia Geoph. et geod. **26** (1982) 405—412.

[10] Laštovičková, M.: Long-Time Measurements of Electrical Conductivity of Some Minerals. Planetary Meeting of European Seismological Commission, Moscow (1984).

[11] Laštovičková, M.: Laboratory Measurements of Electrical Properties of Rocks and Minerals. Geophysical Surveys **6** (1983) 201—213.

Evidence of dissociation kinetics of siderites in their electrical and magnetic parameters

E. I. PARKHOMENKO, K. N. DANUKALOV

Natural carbonates are formed as the result of various processes, e.g. sedimentation under reduction conditions or hydrothermal metamorphism of ultrabasic olivine igneous rocks. Further change of physico-chemical conditions, such as temperature, pressure, O_2-concentration, may cause decomposition of carbonates and the formation of metal oxides and carbon dioxide. In this respect, the reaction of siderite dissociation is interesting. This reaction occurs at fairly low temperatures, whereas its products, the metal oxides maghemite, magnetite, hematite, magnesioferrite, are ore minerals with high conductivity and ferromagnetic properties, which make them useful for solution of paleomagnetic problems. The purpose of this paper is to reveal the effect of pressure on the dissociation kinetics of siderites at different temperatures according to the changes occurring during this process in electrical resistivity and residual magnetic characteristics, i.e., magnetic susceptibility, residual saturation magnetisation, destroying field. The analysed samples of siderites were collected on the Kozhim River (Polar Urals), and from the Bakal (Southern Urals), Karazhal (Kazakhstan) ore deposits. The Karazhal siderites are macrocrystalline, almost without impurities (less than 0.1% of hematite and hydrous ferric oxide); the Bakal siderites are composed of aggregates of large crystallites with 8% of MgO in solid solution; the Kozhim siderites are fine-grained lenslike concretions in association with silicates. The basic impurities are Ca, Al, Mg. The electrical resistivity ϱ was measured by the terraohmmeter E6-13 in isobaric and isothermic regimes and under long-term action of fixed P and T. The temperature interval was 20–600 °C and the pressure interval was 0.1–2.0 GPa. Prior to the experiment, the samples were exposed to 100 °C for 1–2 hours to remove the absorption water. High quasi-hydrostatic pressures were achieved in the cylinder-piston type chamber. Pyrophyllite and fluoroplast were used as the pressure transmitting medium.

The major results of the experiments are described below [1, 2].

Fig. 1 shows the electrical resistivity isobars of the Karazhal and Bakal siderites at $P = 0.1$ GPa. The dissociation of siderites is accompanied by a sudden fall of electrical resistivity by 3–6 orders which is attributed to the formation of ferric oxides, mostly maghemite and magnetite whose conductive properties are determined by the electron exchange mechanism. By comparing the run of temperature ϱ with the results of thermoweight and differential thermal analyses we find that the intensive change of resistivity forestalls the dissociation process. The plot shows that the presence of 8% of Mg in solid solution of siderite-magnesite increases the temperature at which the dissociation pro-

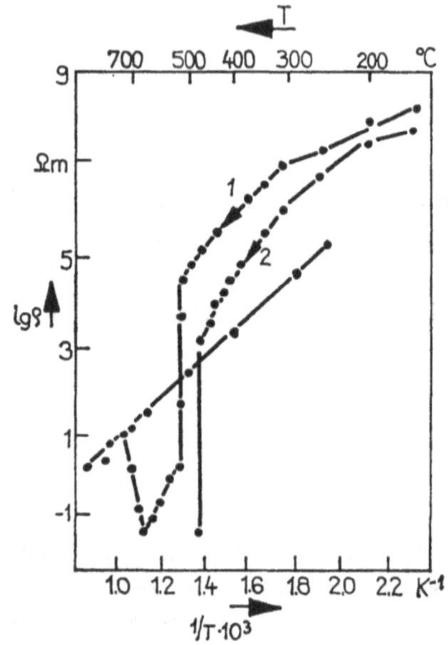

Fig. 1. Isobars of electrical resistivity of the Karazhal (*2*), Bakal (*1*) siderites at $P = 0.1$ GPa.

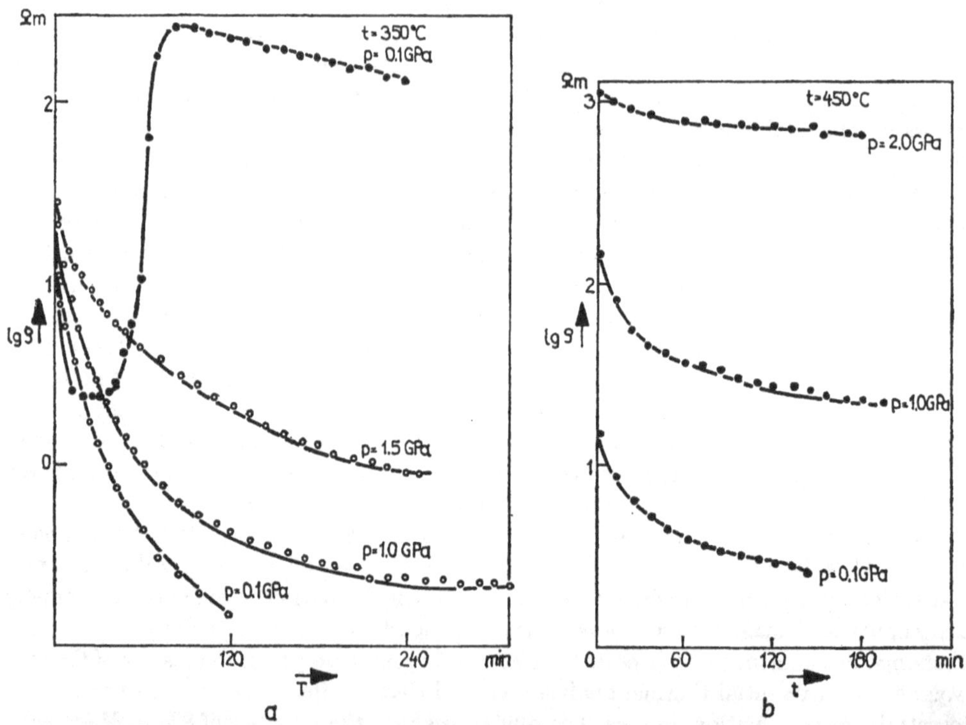

Fig. 2. Curves of dependence of electrical resistivity of the Karazhal (a) and Kozhim (b) siderites on the time of exposure at fixed temperatures and pressures.

cess starts by more than 100°. The presence of magnesium likewise hinders further oxidation of dissociation products to hematite, which fact is reliably fixed from magnetic characteristics and confirmed by data in literature [3].

At temperatures of the decarbonatisation process, the isotherms reveal an irreversible increase of electrical conductivity with pressure, which is partly ascribed to the time effect.

As evidenced by the dependence of electrical resistivity of the Karazhal and Kozhim siderites on the time of exposure at fixed pressures and temperatures and by somewhat lower temperatures of intensive dissociation (Fig. 2), the increase of pressure considerably slows down the dissociation process and entirely suspends further oxidation of reaction products of the pure Karazhal siderite to hematite. The interval of intensive dissociation of the Bakal siderite, containing 8% of Mg in solid solution, is much shorter and the slowing down effect of pressure on this process is less.

The comparison of the thermomagnetograms of dissociation products of the Bakal siderite at pressures 0.1 and 1.5 GPa reveals that the set of blocking temperatures for samples, which experienced higher pressures, is shifted toward high temperatures (Fig. 3). With the increase of pressure the changes in the $Irs(T)$ curves of the Karazhal and Kozhim siderites are similar.

As shown in Table 1, the increase of pressure at the same temperature of the dissociation process controls the increase of the destructing field of saturation of the reaction products. The results of the experiments imply that the increase of pressure hinders the growth of the grains of the reaction products, because at higher pressures more magnetically rigid and, consequently, smaller particles are formed.

The data in the Table also confirm the conclusion about the slowing down of the dissociation process with the increase of pressure, because in this case the destructing field $H'cs$ is the characteristic of magnetic rigidity, whereas magnetic susceptibility $\varkappa$ and residual magnetisation of saturation I_{vs} characterise the amount of newly formed meghamite and magnetite.

It was also established that an intensive reduction of ϱ is observed during the first loading cycle reaching 1.5 GPa in the isothermal regime at temperatures considerably lower than the temperature at the beginning of decarbonatisation; this intensive reduction still continues after removal of pressure. The typical isotherms of electrical resistivity of the Bakal siderite are given in Fig. 4. The isotherms of the first loading were recorded in the range from room temperature to 250 °C at 25° intervals. The cumulative changes of electrical resistivity reach 1.5—4.5 orders. The maximal effect occurs in the temperature interval 100—125 °C. Repeated loading cycles in the isothermal regime do not cause anomaly large changes of ϱ. The general pattern of isotherms of the first loading under low temperatures of the Karazhal and Kozhim siderites is the same as that of the Bakal siderite. When after the first loading cycle the sample is exposed at fixed temperature and under pressure of 0.1 GPa, the resistivity greatly increases, but the initial value is not reached in that process. The resistivity relaxation curves after the first loading are shown in Fig. 5; they indicate that the relaxation process is quicker under higher temperatures. We may suppose, that the mechanism of the effect of sudden change of ϱ of siderite under the first loading and low temperatures is related either to the formation of the ferromagnetic phase of partial dissociation during cracking or to the water in the closed pores, which initially forms the current transmitting channels and then evaporates through newly formed cracks.

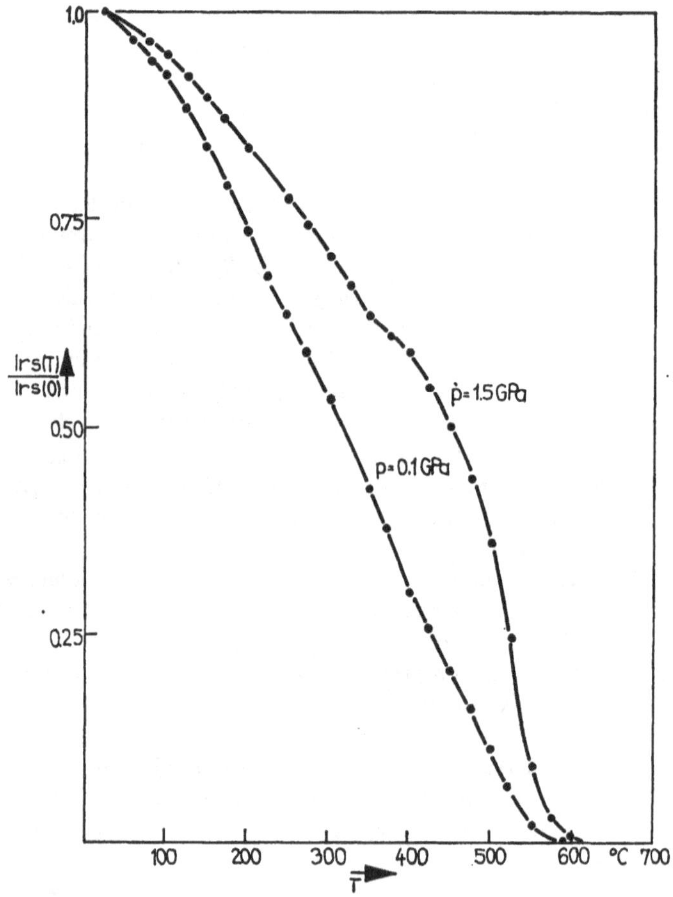

Fig. 3. Thermomagnetograms of dissociation products of the Bakal siderite dissociation which occurred at different pressures.

Table 1. Magnetic properties of siderites dissociation products

Deposit, nos of samples and conditions of experiment	$\varkappa$ (conv. unit)	I_{vs} (10^{-4} T)	H_D (10^2 A^{m-1})
Kozhim R., N 5, exposure at 450°C and 0.1 GPa for 2.5 hours	1743	0.452	374
Kozhim R., N 1, exposure at 450°C and 1.0 GPa for 3h 12 min	241	0.147	726
Kozhim R., N 2, exposure at 450°C and 2.0 GPa for 3 hours	122	0.035	847
Karazhal, N 6, exposure at 350°C and 0.1 GPa for 4 hours	3772	0.928	261
Karazhal, N 5, exposure at 350°C and 1.5 GPa for 2h 24 min	2456	0.591	297

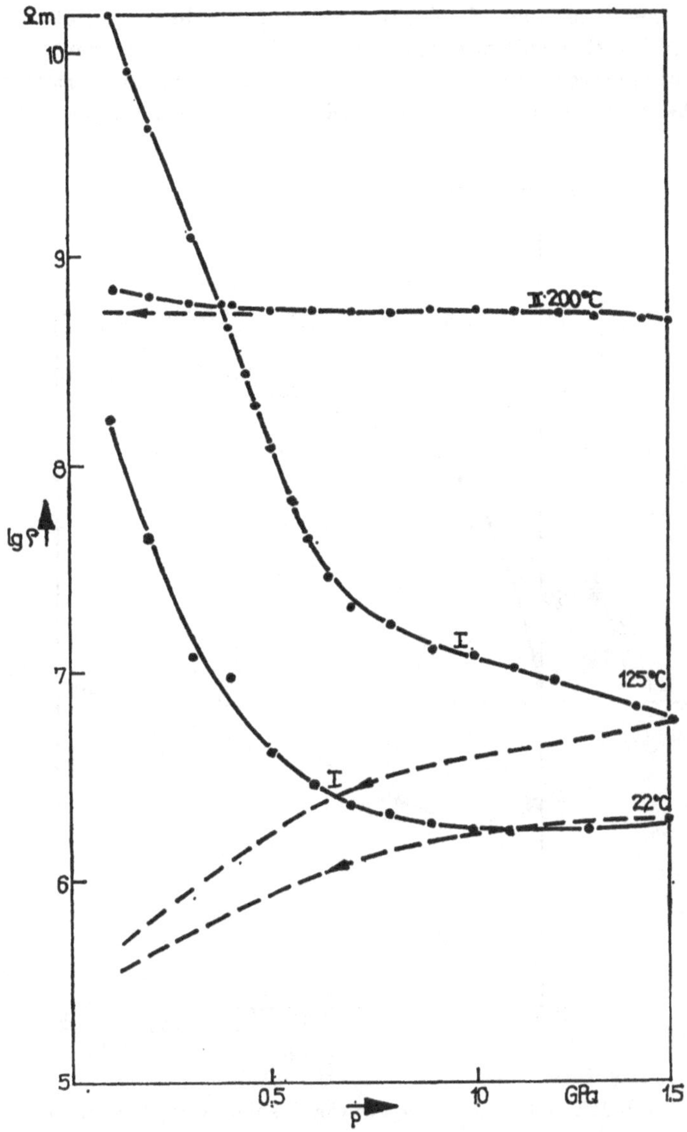

Fig. 4. The electrical resistivity isotherms at initial loading of the Bakal siderite (the isotherm of repeated loading is given for comparison).

Several experiments were conducted to better define this mechanism. The magnetic properties of siderites after the first loading cycle change regularly. In particular, I_{rs} increases 2—50 times depending on the quantity of ferromagnetic impurity in the initial sample, whereas the changes of H'cs indicate the appearance of a new magnetically soft phase. The isotherms of the first loading of siderite in the closed system (fluoroplast) are similar to the curves in Fig. 4, but the cumulative changes of ϱ are less. In the process of the first loading of calcite and magnesite samples, whose crystalline structure resembles that of siderite, the sudden reduction of electrical resistivity is not observed. All this evidence is in favour of the first mechanism of effect of sudden change of ϱ of siderites during the first loading, which means partial dissociation of siderite on the surface of newly formed cracks even at room temperature.

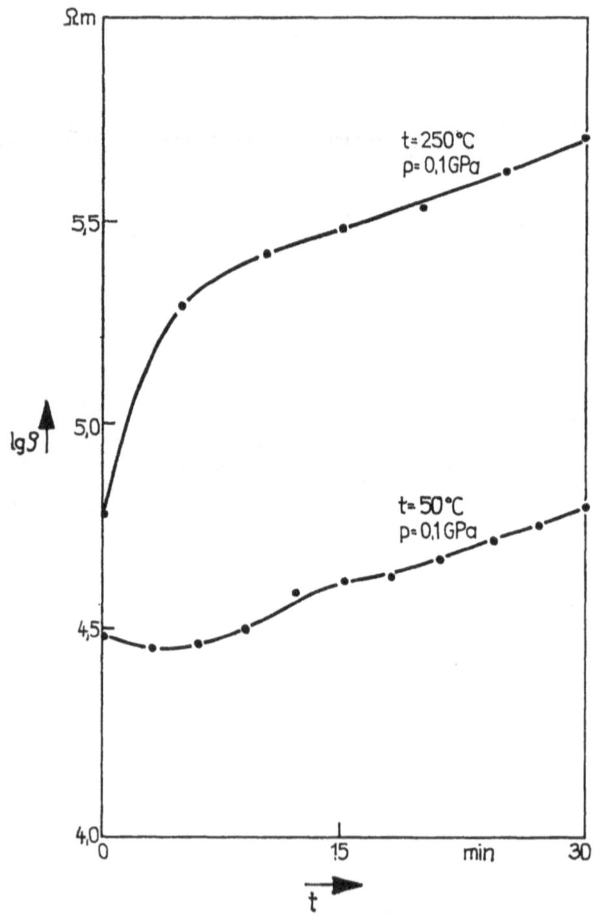

Fig. 5. Resistivity relaxation curves of the Bakal siderite after the first loading.

To conclude, the comparison of the electric and magnetic measurements proved to be rather effective for the study of the decarbonatisation of siderites, which is important both for ore formation problems and for the solution of certain problems of rock magnetism. Moreover, there is the possibility that the thermochemically active rocks may manifest the effect analogous to partial dissociation of $FeCO_3$ on the surface of the newly formed cracks at low temperatures.

References

[1] PARKHOMENKO, E. I.; MKRTCHIAN, S. A.: The chemical composition as the determining factor of electrical resistivity of minerals under high pressures (T — 20 kbar) and temperatures (200—650 °C) (in Russian). Izvestia AN SSSR, Fizika Zemli (1974) 12, 46—58.

[2] PARKHOMENKO, E. I.: Time and pressure effect on electrical properties of siderites at high pressures (in Russian). In: Physical properties of rocks under high pressures and temperatures. Erevan, (1985).

[3] BAGIN, V. I.; GENDLER, T. S.; KUZMIN, R. N.; RYBAK, R. N.: The study of magnetic properties and the Mössbauer effect at the temperatures of siderite transformation (in Russian). Izvestia AN SSSR, Fizika Zemli 11 (1971).

Systems for measurement of rock electrical conductivity in samples of cylindrical form

ANDRZEJ PRAŁAT

Introduction

The most often determined electrical parameter of the rocks is their resistivity or its inverse — the conductivity. It is determined for investigation of deeper strata of the lithosphere and on the measurement of its changes majority of electrical methods of the applied geophysics are based. Apart from that, the resistivity also influences considerably the conditions of the propagation of the electromagnetic waves in the Earth or in its vicinity.

The conductivity (as this quantity will be considered in the farther part) may be determined in conditions of depositions (the field measurements) or by measuring the samples (laboratory measurements). The results of measurements of the conductivity obtained by measuring the samples may differ considerably from the real values of the conductivity in situ. This difference is caused by the pressure, temperature or humidity which are others than those in the deposits. In case of a medium of high heterogeneity, additional errors may be caused also by an insufficient of the sample. However in series of cases, the measurements made on the samples cannot be substituted. One example may be the measurement of samples taken from the bottom of the sea, from the moon or even the investigation of rocks in raised thermobaric regimes. The majority of the investigated samples is a drill core and it has the shape of a cylinder of irregular ends. The present paper is a continuation of the paper [6] and it relates to the comparison of the measuring configurations and also to the obtained results when measuring the cylindrical samples in an electrical field, the height of which samples is several times greater than their diameter.

Presentation of Measuring Configurations

Two-electrode Configuration

In the Fig. 1, two-electrode configurations for the measurement of the conductivity on samples in form of a drill core, are presented. By determining the resistance R (by measuring with a direct or alternating current) between the electrodes, one may calculate the conductivity of the sample from the dependence:

$$\sigma = \frac{K}{R}, \tag{1}$$

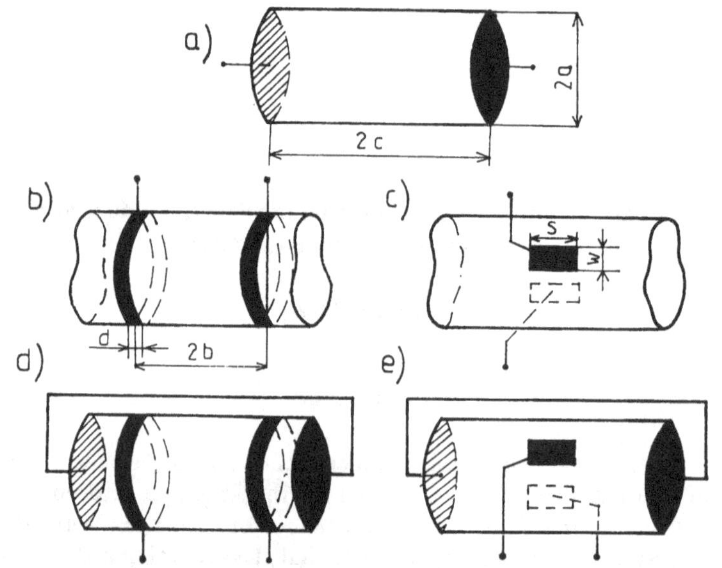

Fig. 1. Two electrode measuring configurations for cylindrical samples
a) with electrodes on the ends, b) with ring electrodes on the surface of the sample, c) with band electrodes on the surface of the sample, d) with measuring ring electrodes on the surface of the sample and with additional electrodes on the ends of an equal potential, e) with measuring band electrodes on the surface of the sample and with additional electrodes on the ends of an equal potential.

where K [l/m], is a coefficient depending on the sample dimensious as well as on the configuration, dimensions and position of the electrodes. For each configuration presented in Fig. 1, the coefficient K should be determined separately, assuming that the sample has an isotropic and homogeneous structure. One must underline here that the resistance R was measured together with the resistance of the electrodes (this is the electrode-sample transition resistance), which in consequences causes that the conductivity determined from the dependence (1), is lower than the real conductivity of the sample. This effect is common for all two-electrode configurations and it is their basic weak point. The evaluation of the quantity of this effect is presented below when measuring in a three-electrode configuration.

In the Fig. 1a, a generally known measuring configuration is presented, in which the sample has a form of a right, circular cylinder and the electrodes are placed at its ends. For this case the coefficient K may be determined in an elementary way and for denotation as in figure it is:

$$K_0 = \frac{2 \cdot c}{\pi a^2}. \tag{2}$$

To realize the above presented configuration, the ends of the drill core segment should be earlier submitted to the working out what is necessary in case of the two following configurations. In the first of them (Fig. 1b), the measuring configuration are two ring electrodes located on the sample. In this case the determination of the coefficient K is more complicated: it is necessary to determine the distribution of the voltage on the

Table 1. Dependence $K_1 = f(2b)$ calculated from Fig. 1.b. for the dimensions of electrodes and sample presented below

$2c = 170$		$a = 23$	$d = 10$
$2b = 20$	to	150	width of step 10
$2b = 20$			$K_1 = 16 \cdot 08$
$2b = 30$			$K_1 = 22 \cdot 22$
$2b = 40$			$K_1 = 28 \cdot 26$
$2b = 50$			$K_1 = 34 \cdot 28$
$2b = 60$			$K_1 = 40 \cdot 3$
$2b = 70$			$K_1 = 46 \cdot 43$
$2b = 80$			$K_1 = 52 \cdot 44$
$2b = 90$			$K_1 = 58 \cdot 46$
$2b = 100$			$K_1 = 64 \cdot 48$
$2b = 110$			$K_1 = 70 \cdot 49$
$2b = 120$			$K_1 = 76 \cdot 51$
$2b = 130$			$K_1 = 82 \cdot 53$
$2b = 140$			$K_1 = 88 \cdot 57$
$2b = 150$			$K_1 = 94 \cdot 71$

sample from the equation by Laplace [1]. For the electrodes located symmetrically with the distance between their centers $2b$ and width d, the coefficient for the ring electrodes is determined by:

$$K_l = \frac{2}{\pi c} \sum_{r=0}^{\infty} \left(\frac{I_0(k_r a)}{k_r a I_0'(k_r a)} \right) \left(\frac{\sin (k_r d/2)}{k_r d/2} \right) \sin^2 k_r b , \tag{3}$$

where: $k_r = (2r + 1)\, \pi/2c$,

I_0 and I_0' — are a modified Bessel function of zero order and its derivative.

The course of the dependence $K_l = f(a, c)$ was presented in the former paper [6]. To compute easier the coefficient K_l, as well as the coefficients for the remaining two-electrode configurations given below, a program on microcomputer SPECTRUM 48 k, has been made, which enables to calculate the coefficient at the assigned geometrical dimensions of the samples and electrodes and allowing to compute the dependence of this coefficient from one arbitrary dimension. In the Table 1, a print out for the dependence $K_l = f(2b)$ is given as an example, for the dimensions which will be used in a further part of the paper in the measurements.

A second two-electrode configuration designed for the .measurement of samples of irregular ends is the configuration presented in the Fig. 1c. For two equal band electrodes of a length s and width w, located on the opposite, sites of the sample and centered about in the middle of the cylinder, the coefficient is defined [1]:

$$K_3 = \frac{4}{\pi^2 a} \sum_{r=1}^{\infty} \frac{1}{r} \sum_{n=1}^{\infty'} \frac{\sin\left(\frac{r\pi s}{2c}\right)}{\frac{r\pi s}{2c}} \frac{\sin\left(\frac{nw}{2a}\right)}{\frac{nw}{2a}} \frac{I_n\left(\frac{r\pi a}{c}\right)}{I_n'\left(\frac{r\pi a}{c}\right)} + \frac{4a}{\pi c w} \sum_{n=1}^{\infty'} \frac{1}{n^2} \sin\left(\frac{nw}{2a}\right), \tag{4}$$

where: I_n and I_n' — are a modified Bessel function of n order and its derivative. The prim on the summation means that only the odd n values should be used.

The measuring configurations presented in the Fig. 1d, and e are designed to investigate the rock conductivity in a thermobaric regime. When realizing the pressure on the sample ends, the steel structure of the press causes that these ends are on the same potential. Hence results the necessity for placeing the measuring electrodes on the surface of the sample.

The condition of equal potentials of both samples ends changes the solution of Laplace equation and hence the previously given coefficients for the same configurations of the measuring electrodes on the surface of the cylinder, but however without electrodes connected between themselves on the ends are not important. For the measuring configurations with additional electrodes on the ends, the determination of the coefficients K may be made on the basis [2]. In case of ring electrodes, located symmetrically at a $2b$ distance between themselves and of a width d (Fig. 1d) this coefficient is given by:

$$K_2 = \frac{4}{dac\pi} \sum_{r=1}^{\infty} \frac{I_0\left(\frac{r\pi a}{c}\right)}{I_0'\left(\frac{r\pi a}{c}\right)} \frac{\sin(k_r d)\sin^2(k_r b)}{k_r^2}, \tag{5}$$

where: $k_r = r\pi/c$.

Likewise by an other dependence than previously it is expressed the coefficient in case of band electrodes of a length s and width w located on the surface of the samples symmetrically on the length and on the parameter:

$$K_4 = \frac{64c}{\pi^2 sw} \sum_{r=0}^{\infty} \sum_{n=1}^{\infty'} \frac{\sin\left(\frac{nw}{2a}\right)\sin\left(\frac{(2r+1)\pi s}{4c}\right)}{(2r+1)^2 n} \frac{I_n\left(\frac{(2r+1)\pi a}{2c}\right)}{I_n'\left(\frac{(2r+1)\pi a}{2c}\right)} \tag{6}$$

the same designations as in previous dependence being applied.

In spite of enough complicated mathematically dependences on the coefficients the sum of modified quotients on n-order Bessel's function and of their derivatives, these dependences can be calculated with an assumed accuracy better than 1% according to the program elaborated to this end.

In case of a measuring configuration in form of two ring electrodes presented in the Fig. 1b, the dependence for the calculation of the coefficient may also be determined with an approximation, however, with an accuracy better than +10% in comparison to the value calculated from 3. The approximated dependence has the form:

$$K_5 = \frac{b}{\pi ad} \tag{7}$$

from which it results that to calculate the conductivity in a configuration as in the Fig. 1b, one may use an equivalent sample in form of a rectangular prism with its height equal to the distance between the electrodes $2b$ and of an end surface equal to the electrode surface: $2\pi ad$.

Four-electrode Configuration

The advantage of these configurations in comparison to two electrodes ones, is the elimination of electrode resistance influence on the results of measurements, if only the measurement of the voltage is carried out in a configuration of a high input resistance; what it is possible to be achieved in the whole range of measured conductivities for rocks in the present state of the measuring technique.

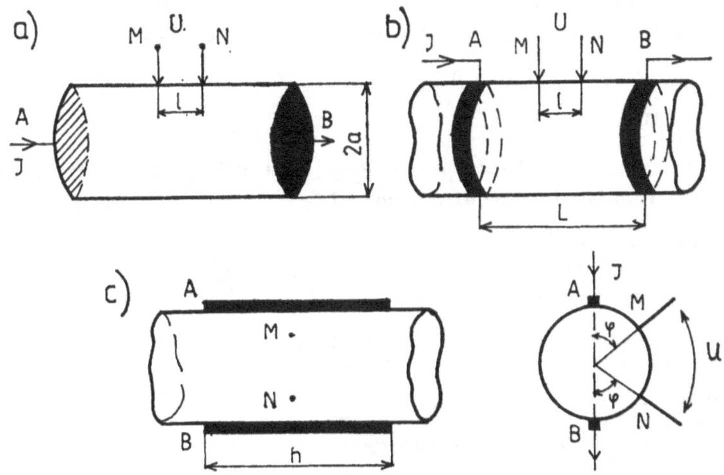

Fig. 2. Four-electrode measuring configuration for cylindrical samples
a) with current electrodes on the ends and voltage point electrodes on the surface of the samples, b) with current ring and voltage point electrodes on the surface of the samples, c) with electrodes on the surface of the samples: linear current and point voltage.

In the Fig. 2, three four electrode measuring configurations are presented, the two of them appearing from the development of the basic configurations presented in the Fig. 2a, quoted in the monograph on electrical parameters of the rocks [5], in case of which the conductivity is expressed from the dependence:

$$\sigma = \frac{Il}{U\pi a^2}, \tag{8}$$

where: I — the current flowing through the sample, through the current electrodes A, B

U — the difference of the potentials measured between the point voltage electrodes M, N

a — the radius of the sample

l — the distance between the voltage electrodes.

The above presented measuring configuration can be realized only in case of samples in form of right circular cylinders. In case of samples of drill core segments, we may apply a modification of the previous configuration, based on locating the current electrodes in form of rings on the surface of the sample (Fig. 2b). However in order to maintain the condition of an electric field homogeneity, it is required to have a large distances between the current electrodes L in comparison to the distance l between the voltage electrodes. When $L \geqq 5l$, the measured conductivity in the configuration as in Fig. 2b may be determined from [8].

Another modification of the basis four electrode configuration has proposed Parasnis [3] (Fig. 2c). In this configuration linear current electrodes gave been applied, which were located on the surface of the samples parallel to its axis and displaced with a π angle, while the point voltage electrodes are also on the surface of the sample, symmetrically to the voltage electrodes. In this case, the conductivity is espressed by:

$$\sigma = \frac{kI}{hU}, \tag{9}$$

where: I and U — are the current and voltage as in other four electrode methods.
h — length of the linear current electrode
$k = \frac{l}{\pi} \ln \frac{1 + \cos \varphi}{l - \cos \varphi}$
φ — angle marked in the Fig. 2c.

If $\varphi = 60°$, then the dependence (9) is brought to

$$\sigma = 0.35 \frac{I}{hU}. \tag{10}$$

Three-electrode Configuration

Independently on the presented measuring configuration, two and four electrode configurations, it is proposed a three-electrode configuration presented in the Fig. 3 which enables to determine the conductivity of the sample with elimination of the electrode

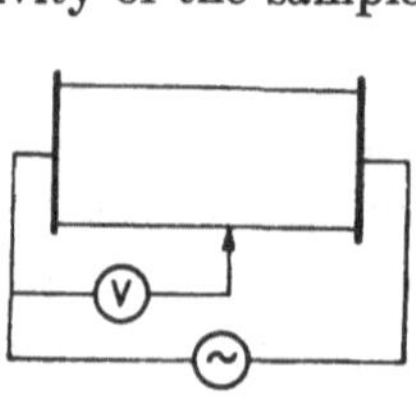

Fig. 3. Three-electrode measuring configuration for cylindrical samples

resistance influence. An incontestable advantage of the three electrode configuration is the possibility to state the heterogeneity of the sample structure, if of course this heterogeneity causes the change of the conductivity. In this configuration, the source of the current and the voltmeter have one terminal each on the same potential (this point can be earthed) what is a basic advantage in comparison to the four-electrode configurations.

The measurement in a three electrode configuration are based on the measurement of the voltage distribution along the sample with a constant current flowing through it. The way to establish the conductivity and the evaluation of the heterogeneity will be presented in the example as below.

The Measurements

Description of Measurements

To carry out the measurements enabling to compare the above presented measuring configurations, samples simulating with their electrical properties the rock samples have been prepared. The samples of the rock may change their humidity, what causes the change of their conductivity with time. Prepared from the mixture of graphite and epoxy resisn, these samples are deprived of the said weak point.

All the electrodes (except the point ones) are made of copper foil of 0.05 mm thickness. The point electrodes were the steel points.

The measurements have been carried out at a frequency 1 kc/s and it was stated that identical results were also obtained at *dc*.

Result of Measurements

To evaluate the homogeneity of the samples, measurements in a three electrode configuration have been carried out and their results for the sample 1 are presented in the Fig. 4. From the presented run of the voltage, measured along the sample, it results that the sample is characterized by a certain heterogeneity, what is observed from the divergence between the curve of the dependence of this voltage (line 1) and its approximation — straight (line 2). This heterogeneity results from an irregular distribution of the graphite during the preparation of the sample. From the results of measurements in the three electrode configuration, one may evaluate also the quantity of the electrode resistance influence on the measured value of the sample conductivity. For the case presented in the Fig. 4, it is easy to determine, that the relation of the real sample conductivity to the measured conductivity in a two electrode configuration is expressed by the quantity $U/(U_2 - U_1)$. However, to illustrate this effect, one may underline,

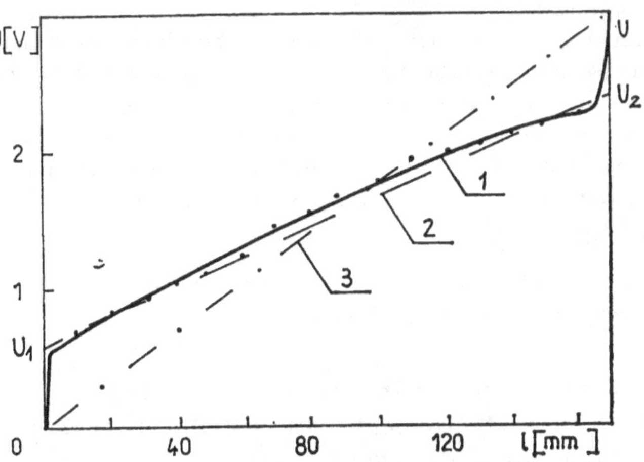

Fig. 4. Line of the voltage measured along the sample No. 1 in a three-electrode configuration (*1*) and its approximation by straight line (*2*). The line (*3*) represents the run for an ideal sample without the effect of the electrode resistance.

Table 2. Results of conductivity measurements of two samples carried out in particular configuration.
Results expressed in percentage in comparison to results obtained by four-electrode configuration (Fig. 2.a.)

measurements system	sample No. 1	sample No. 2
Fig. 1.a.	−25	+5
Fig. 1.b.	−11	0
Fig. 1.d.	−6	+6,3
Fig. 1.c.	−13	+5,2
Fig. 1.e.	−20	+3,5
Fig. 1.b.[1]	−3	+6
Fig. 2.a.	0	0
Fig. 2.b.	−3	−3
Fig. 2.c.	−70	−70
Fig. 3.	+4	+8,5

[1] calculated from approximated dependence

that it has been chosen a sample, the influence of which on the electrode resistance be particularly high.

In the Table 2, results of the conductivity measurements for two samples, carried out in the above presented configuration, are put together for comparison. These results have been expressed proportionally in comparison to the value of the conductivity measured in a four electrode configuration as shown in the Fig. 2a. These values for the sample Nr. 1 and Nr. 2 are 0.98 mS/m and 96 mS/m respectively. When determining the conductivity in particular configurations, the above presented dependences have been used.

Range of Measurement Frequency

When determining the conductivity on each of the above presented measuring configurations, it is assumed that the distribution of the potential V on the sample may be determined from the equation of Laplace $\nabla^2 V = 0$. This means that the electrical field should be stationary, i.e. to be produced in principle by a direct current. However, it is assumed [4] that the above presented solution of Laplace equation is correct also for the quasi stationary variable fields of a frequency f, at which the condition

$$ks \leqq 10^{-2} \tag{11}$$

where: k — is a constant of propagation at this frequency,
s — maximum dimension of the sample

is fulfilled.

For the electrical parameters of the conductivity σ, relative electric permeability ε_r and relative magnetic permittivity μ_r, the constant of propagation is given by:

$$k = (\omega^2 \mu_0 \mu_r \varepsilon_0 \varepsilon_r - i\omega\mu_0\sigma)^{1/2}, \tag{12}$$

where: $\omega = 2f\pi$,
ε_0 and μ_0 — electrical and magnetical permeability for the vacuum.

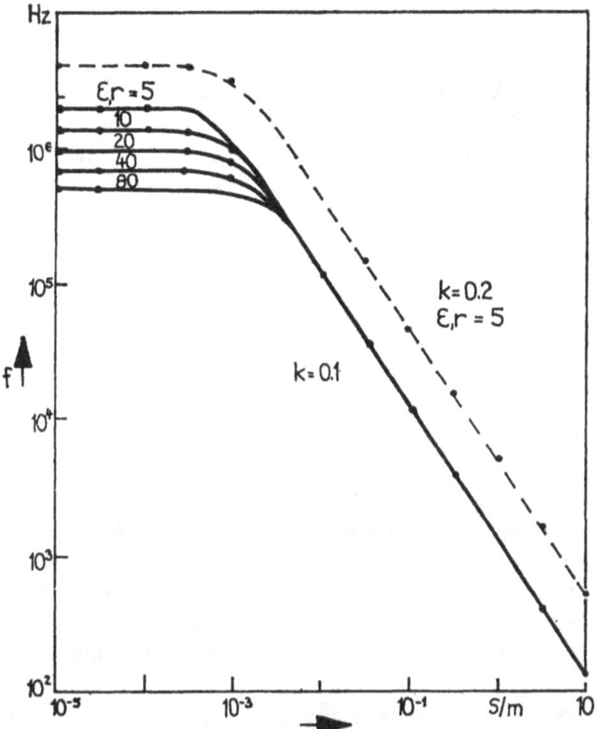

Fig. 5. Dependence of the frequency on electrical parameters of the medium for the propagation constant 0.1 and 0.2.

The dependence [12] may be solved in consideration of the frequency:

$$f = \frac{\sigma}{2\sqrt{2\pi\varepsilon_r\varepsilon_0}} \sqrt{\sqrt{1 + \left(\frac{2\varepsilon_r\varepsilon_0 k^2}{\mu_r\mu_0\sigma^2}\right)^2} - 1}. \tag{13}$$

In the Fig. 5, the frequency in function of electrical conductivity and permeability calculated from (13) is presented, assuming that $\mu = 1$ for $k = 0.1$, i.e. the maximum value of the frequency, at which the condition (11) is fulfilled for a maximum dimension of the sample $s = 0.1$ m. From this diagram one may evaluate that the condition of the quasistationary for the samples and their conductivities used in the measurements is fulfilled.

However, it should be underlined, that the above presented consideration is related only to the sample itself, while in a real measuring configuration, it is necessary to take also into account the leads, in consequence of which the measurement frequency should be lower than it results from the dependence (13). This relates particularly to the four-electrode configurations, in which the inductivity of the leads and the coupling between them should be taken into consideration, by determining the conductivity at high frequencies. The obtention of quantitative results, requires a separate analysis of the problem.

Recapitulation

Nine configurations for the measurement of cylindrical samples are presented above and also particular configurations have different purposes, as e.g. to measure the segments of a core of irregular ends or to measure in the pressure chambers. The comparative measurements carried out on samples prepared specially for this purpose, simulating the electrical properties of the rock samples, enable to obtain series of conclusions.

In one of the verified configurations: in a four-electrode configuration with linear current-electrodes, over three times lower value of the conductivity has been obtained. One may think that the coefficient in the dependence (9) has been inappropriately assumed. An analytical determination of this coefficient will be the subject of a separate paper.

The samples, the result of measurement of which have been presented by way of example, differ by their conductivity approximately hundred times. For a sample of a smaller conductivity (No. 1), the divergence between the results of measurements obtained in particular measuring configurations is visibly higher and it is over 30%; lower conductivity values being observed in the two-electrode configurations.

For a sample of a higher conductivity (No. 2) the divergence between the values of the conductivity determined in particular configurations is less than 12% and the values determined in the four-electrode configurations are lower than those in two-electrodes.

For both the samples, the conductivity determined from the measurements in a three-electrode configuration is higher.

During the measurements, it was stated that the used samples, show a certain heterogeneity (e.g. the sample No. 1, particularly of the end surfaces) and in several measuring configurations, the density of the current line is different in different parts of the sample (i.e. it measures as if other parts of the sample). These facts as well as the small number of the investigated samples do not allow to carry out a rigorous comparison as particular measuring configurations with respect to their measurement accuracy but only to evaluate with which divergencies in the result of measurements one may expect to find in the presented measuring configurations.

References

[1] Lytle, R. J.; Duba, A. G.; Wilows, J. L.: Alternative metods for determing the electrical conductivity of core samples, Rev. Sci. Instrum. 1979. **50** (1979) 5, 611–615.

[2] Lytle, R. J.: Alternative methods for determining the electrical coductivity of core samples II, Rev. Sci. Instrum. 1983, **54** (1983) 7, 871–874.

[3] Parasnis, D. S.: Mining Geophysics, Elsevier, Amsterdam (1973).

[4] Plonsey, R.; Colin, R. E.: Principles and Applications of Electromagnetic Fields. McGraw-Hill, New York (1961).

[5] Parkomenko, E. I.: Elektriceskije svojstva gornyk porod (in Russian). Nauka, Moskva (1965).

[6] Prałat, A.: Skema dla izmerenija elektriceskovo soprotovlenija obrascov porod (in Russian). Tezisy dokładov Tematiceskovo Simpoziuma i Rabocevo Sovescanije Projekta, 3 KAPG, Miskolc (1983) 39–41.

Permeability from electrical properties — a comparative method for the determination of ultra-low permeability under pressure

D. Freund

1. Introduction

Relatively impermeable rock formations are important to subterranean storage of toxic and/or high-level radioactive waste, to earthquake prediction and to isolation of hot-dry-rock geothermal systems and natural gas reservoirs. While for the determination of permeabilities $> 10^{-4}$ md the conventional steady-state flow techniques are used — for ultralow permeabilities in the range of 10^{-4} to 10^{-8} md the transient pulse technique TPT is preferred. This method developed by Brace et al. [1968] and modified and improved by Zoback [1975], Trimmer et al. [1980], Trimmer [1981, 1982], Yamada [1981], Lin [1982] and Lin et al. [1986] makes use of transient fluid flow in porous geological materials. The decay of a small incremental change of pressure (pressure gradient diffusion) is observed. High-pressure permeability measurements for simulation of conditions of the Earth's uppermost crust can be carried out — in the most simple case of a stress state — by experiments under hydrostatic loading.

The isolation of natural gas fields by very low porosity (argilliceous) cap rocks depends among other properties on their ultra low permeability. For a set of such rocks it will be tried to get information about the behaviour of their permeability reduction resulting from electrical measurement under pressure.

2. Theoretical relations between porosity, permeability and electrical conductivity

On the basis of Darcy's law for fluid flow and Archie's empirical law permeability and electrical conductivity are related by [Brace et al., 1965, Brace et al., 1968, Brace, 1977]

$$k = (m^2/b)\, F^{-1,5}, \tag{1}$$

where k is permeability, m is the hydraulic radius (volume of interconnected pores referred to surface area), b is a shape factor, and F is the formation factor (the ratio of the conductivity of an electrolytic saturant to the conductivity of the fully-saturated sample: σ_e/σ_r).

A similar connection is obtained by Walsh and Brace [1984] for experiments under hydrostatic pressure for various granites and other low porosity crystalline rocks. The same authors refer to the similarities of the diffusion equations for the volumetric flow rate q under a pressure gradient ∇p given by

$$(q/A) = -(k/\mu)\, \nabla p \tag{2}$$

(where A is the cross-section, k is the permeability, μ is the viscosity of the pore fluid) and the flow of current under a potential gradient ∇v which can be written in an analogous manner

$$J = -c \cdot \nabla v, \tag{3}$$

where J is the current flux density, and c is the conductivity.

The version of Archie's law

$$\sigma_r = a \cdot \sigma_e \Phi^m \tag{4}$$

combines the conductivity with porosity Φ and is consistent in this power-law form with the empirical Kozeny-Carman equation

$$k = c \frac{\Phi^3}{So^2}. \tag{5}$$

Here So is the specific surface area and c is an empirical constant ≈ 0.2 [WONG, KOPLIK and TOMANIC, 1984]. The capillary bundle model expressed in the Kozeny-Carman equation also represents the starting point of a network model after SCHOPPER [1966] and relates permeability to formation factor and porosity.

The basic relation described by PAPE, RIEPE and SCHOPPER [1981], [1982], [1985]

$$K = \frac{1}{F} \cdot \frac{1}{2S_{\text{por}}^{*2}} \tag{6}$$

(where S_{por}^{*} is the capillary surface) is identical with

$$m = (bkF)^{1/2} \tag{7}$$

from WALSH and BRACE [1984], if $b = 2$ (for circular tubes) and

$$m = \frac{1}{S_{\text{por}}^{*}}. \tag{8}$$

For sedimentary rocks with permeabilities ranging from 10^4 to 10^{-2} md an empirical relation is shown by the so-called "first Paris equation"

$$k = \frac{1}{F} \cdot 475 \cdot \frac{1}{S_{\text{por}}^{\text{BET}3.1085}} \tag{9}$$

[PAPE et al., 1981]. $S_{\text{por}}^{\text{BET}}$ is the BET-surface measured in the laboratory by nitrogen adsorption, referred to the pore space.

3. Sample description and experimental procedures

In Table 1 mineral content determined by quantitative X ray analysis of the fine-grained sandstones are given. All samples have a very low porosity and are possible representatives of intraformation cap rocks. They are cored from a deep bore hole and

Table 1. Sample Characterisation Data

Sample	Mineral Content (%) by X Ray Method						
	qz	pl	K-fsp	anh	calc	dol	clay
1	48	6	2	4	5	11	24
9	43	7	3	14	4	11	18
11	38	7	2	9	8	2	34
14	54	5	4	12	2	6	17
18	42	7	2	19	6	—	24
21	47	8	3	22	5	4	11

Table 2. Air Permeability K as a Function of Effective Pressure P_{eff}

Sample	k (md) at P_{eff} (MPa)			
	5	20	40	80 MPa
1	$1.84 \cdot 10^{-3}$	$3.5 \cdot 10^{-4}$	$7.6 \cdot 10^{-5}$	$1.3 \cdot 10^{-5}$
9	$1.3 \cdot 10^{-3}$	$5.0 \cdot 10^{-5}$	$8.0 \cdot 10^{-6}$	$3.0 \cdot 10^{-6}$
11	$5.4 \cdot 10^{-3}$	$2.4 \cdot 10^{-4}$	$5.6 \cdot 10^{-5}$	$1.0 \cdot 10^{-5}$
14	$1.23 \cdot 10^{-2}$	$9.6 \cdot 10^{-4}$	$1.5 \cdot 10^{-4}$	$3.1 \cdot 10^{-5}$
18	$5.9 \cdot 10^{-4}$	$3.5 \cdot 10^{-5}$	$5.0 \cdot 10^{-6}$	—
21	$2.35 \cdot 10^{-3}$	$3.5 \cdot 10^{-4}$	$5.2 \cdot 10^{-5}$	$2.0 \cdot 10^{-6}$

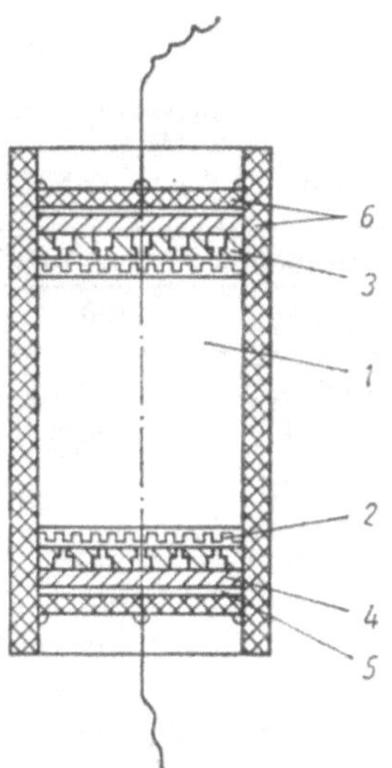

Fig. 1. Sample arrangement for conductivity measurements under drained conditions. (Description see text.)

prepared as cylindrical specimens ($d = 30$ mm). After careful drying (60 °C; 100 h) the air permeability has been measured perpendicular to the bedding plane up to an effective pressure of 80 MPa. These measurements were carried out by Iffland [1984]. The permeability values at effective pressure of 5, 20, 40 and 80 MPa are shown in Table 2. Permeability is reduced about 3 orders of the magnitude range.

Electrical data were measured in a second pressure run after an appropriate relaxing time (≈ 2 months). The sample arrangement illustrated in Fig. 1 [Brace, 1965] was used for the conductance measurements in a common piston-cylinder-apparatus which is driven by a hydraulic ram. At each end of the cylindrical specimen (*1*) there is at first a metallic filter of high porosity (*2*) together with steel reservoirs (*3*) which maintain the drained conditions during the experiments. That means that pore pressure is near atmospheric and the confining pressure is equal to an effective pressure. Next to the steel reservoirs steel discs (*4*) and electrodes (copper-foil) (*5*) are situated. The entire arrangement is encapsulated by a rubber jacket (*6*) for insulation pressure medium from the sample.

The specimens are fully saturated with 0.2 m NaCl by a special method. Pressure steps of the electrical measurements are identical with the low pressure cycle up to 80 MPa of the permeability test. The time interval between the different pressure steps during electrical experiments must be at least equal to or greater than those which are necessary for the detections of permeability data.

4. Results

The results of electrical measurements are listed in Table 3. The data valuated up to pressures of 200 MPa (No. 9 up to 300 MPa) are given as conductivity σ ($\Omega^{-1} \cdot m^{-1}$) and formation factor (dimensionless).

In Fig. 2 (upper part) permeability and formation factor are plotted in a double logarithmic scale. The following straight line according to

$$\log k \propto -r \cdot \log F \tag{10}$$

for all samples is obtained within an unexpected good correlation (correlation coeffi-

Table 3. Results of Electrical Measurements

Sample	Conductivity ($\Omega^{-1} \cdot m^{-1}$) and Formation Factor at Different Pressures (MPa)							
	5	20	40	80	120	160	200	300 MPa
1	$1.22 \cdot 10^{-2}$	$1.02 \cdot 10^{-2}$	$7.37 \cdot 10^{-3}$	$4.72 \cdot 10^{-3}$	$3.83 \cdot 10^{-3}$	$3.26 \cdot 10^{-3}$	$2.89 \cdot 10^{-3}$	—
	155	185	256	400	494	580	653	—
9	$3.29 \cdot 10^{-2}$	$2.06 \cdot 10^{-2}$	$1.30 \cdot 10^{-2}$	$8.74 \cdot 10^{-3}$	—	$5.55 \cdot 10^{-3}$	$5.10 \cdot 10^{-3}$	$4.48 \cdot 10^{-3}$
	58	92	145	216	—	340	375	421
11	$4.05 \cdot 10^{-2}$	$2.70 \cdot 10^{-2}$	$2.12 \cdot 10^{-2}$	$1.66 \cdot 10^{-2}$	$1.52 \cdot 10^{-2}$	$1.39 \cdot 10^{-2}$	$1.35 \cdot 10^{-2}$	—
	61	91	116	148	168	179	182	—
14	$1.63 \cdot 10^{-2}$	$1.24 \cdot 10^{-2}$	$1.00 \cdot 10^{-2}$	$7.66 \cdot 10^{-3}$	$6.69 \cdot 10^{-3}$	$5.46 \cdot 10^{-3}$	$4.49 \cdot 10^{-3}$	—
	133	175	217	283	344	421	483	—
18	$1.65 \cdot 10^{-2}$	$1.31 \cdot 10^{-2}$	$9.37 \cdot 10^{-3}$	$6.47 \cdot 10^{-3}$	$6.02 \cdot 10^{-3}$	$5.62 \cdot 10^{-3}$	$5.17 \cdot 10^{-3}$	—
	127	160	224	325	356	374	406	—
21	$2.99 \cdot 10^{-2}$	$1.94 \cdot 10^{-2}$	$1.08 \cdot 10^{-2}$	$6.30 \cdot 10^{-3}$	—	—	—	—
	63	97	176	297	—	—	—	—

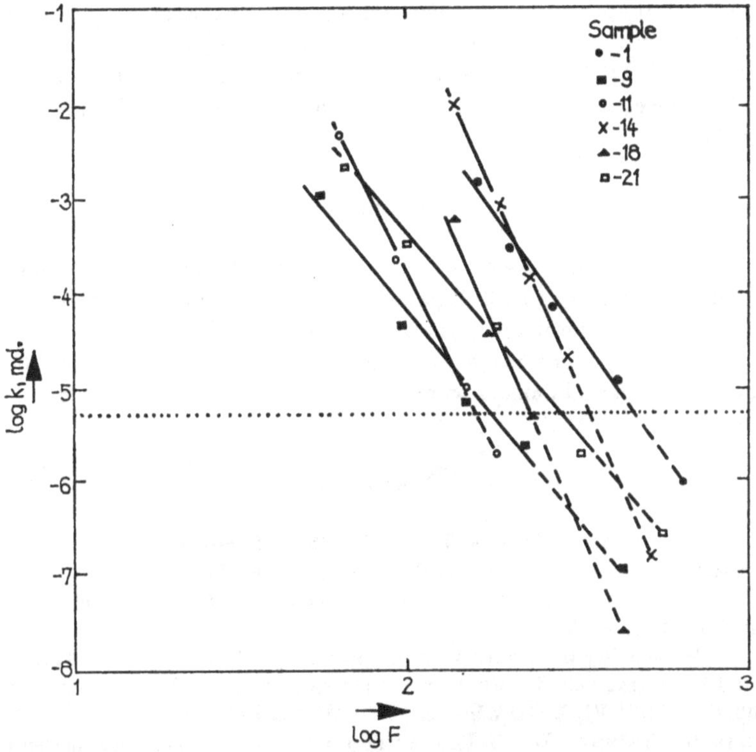

Fig. 2. Relation $k \propto F^{-r}$. Experimental data at $P_{eff} = 5$; 20; 40; 80 MPa and extrapolated (dashed lines) k-values from electrical conductivity (up to 300 MPa). The dotted line indicates the lower limit of the available steady-state flow technique for the permeability determination.

cient averages 0.97). The relation (9) can also be described by

$$k \propto \sigma^r . \tag{11}$$

The exponent r for the different samples is given (together with initial porosities) in Table 4 and depends on pore size and geometrical structure of the pore surface. Further investigations of the inner pore structure are irremiscible.

Table 4. Initial ($P = 5$ MPa) Porosities and Slope r of $k \propto F^{-r}$ in the Pressure Range from 5 to 300 MPa

Sample	Initial Porosity (%)	r
1	1.7	5.0
9	1.2	4.4
11	2.3	7.0
14	2.3	8.5
18	1.5	8.2
21	0.95	4.4

The strong interdependence of permeability and electrical properties in the low pressure range permits to extrapolate ultra low permeability values (dashed lines) into the high-pressure field (Fig. 2, lower part). This result indicates that permeability can be derived from electrical measurements under hydrostatic stress for low-porosity sedimentary rocks, too.

Summary

Gas permeability up to an effective pressure of 80 MPa and electrical data (conductivity, formation factor) up to 200 MPa are determined for a set of cap rocks.

The close connection between permeability ($\log k$) and formation factor ($\log F$) in the lower pressure range, according to $k \propto F^{-r}$, allows an extrapolation of permeability data from electrical measurements to ultra-low values at high pressures.

References

Brace, W. F.; Orange, A. S.; Madden, T. M.: The effect of pressure on the electrical resistivity of water saturated crystalline rocks. J.G.R. **70** (1965) 5669–5678.

Brace, W. F.; Walsh, J. B.; Frangos, W. T.: Permeability of granite under high pressure. J.G.R. **73** (1968) 2225–2236.

Brace, W. F.: Permeability from resistivity and pore shape. J.G.R. **82** (1977) 3343–3349.

Iffland, J.: Ein Beitrag zur Bewertung von Speichern geringer Porosität und Permeabilität. Abschlußbericht 84/81 VD B-GII/53/84 KEE GOMMERN (1984).

Lin, C.; Pirie, G.; Trimmer, D. A.: Low permeability rocks: laboratory measurements and three-dimensional microstructural analysis. J.G.R. **91** (1986) 2173–2181.

Lin, W.: Parametric analyses of the transient method of measuring permeability. J.G.R. **87** (1982) 1055–1060.

Pape, H.; Riepe, L.; Schopper, J. R.: Calculating permeability from surface area measurements 7th Europ. Logging Symp. Trans. Paris 1981.

Pape, H.; Riepe, L.; Schopper, J. R.: A pigeon-hole model for relating permeability to specific surface. Log Analyst **23** (1982) 5–13.

Pape, H.; Riepe, L.; Schopper, J. R.: Permeability of porous rocks derived from internal surface effects. Preprint 47th EAEG Meeting, Budapest, 4.–7. June 1985.

Schopper, J. R.: A theoretical investigation on the formation factor/permeability/porosity relationship using a network model. Geophys. Prospect. **14** (1966) 301–341.

Trimmer, D.; Bonner, B.; Heard, H. C.; Duba, A.: Effect of pressure and stress on water transport in intact and fractured gabbro and granite J.G.R. **85** (1980) 7059–7071.

Trimmer, D.: Design criteria for laboratory measurements of low permeability rocks. Geophys. Res. Lett. **8** (1981) 973–975.

Trimmer, D.: Laboratory measurements of ultralow permeability of geological materials. Rev. Sci. Instr. **53** (1982) 1246–1254.

Walsh, J. B.; Brace, W. F.: The effect of pressure on porosity and the transport properties of rock. J.G.R. **89** (1984) 9425–9431.

Wong, P.; Koplik, J.; Tomanic, J. P.: Conductivity and permeability of rocks. Physical Review B, **30** (1984) 6606–6614.

Yamada, S. E.; Jones, A. H.: A review of a pulse technique for permeability measurements. SPEJ **20** (1980) 357–358.

Zoback, M. D.; Byerlee, J. D.: Permeability and effective stress. Am. Ass. Petrol. Geologists Bull. **59** (1975) 161–175.

CHAPTER III

Thermal Properties

Thermal conductivity of certain rocks under high pressures and temperatures

S. N. Emirov

In the light of the principles of modern geomechanics, the rocks do not represent a continuous medium, but are composed of a series of blocks [1]. Therefore, the problem of study of thermal conductivity of real rocks has two stages: the study of the dependence of thermal conductivity on pressure and temperature, and the study of the dependence of thermal conductivity on block structure.

For experimental study of the effect of hydrostatic pressure up to 350 MPa and temperature (273—423 K) on thermal conductivity of dielectrics and rocks the absolute stationary compensation method was used [2] (Fig. 1). The gradient heater *2* is placed between both samples *1* of the studied material having identical size and cut out from one polycrystalline ingot. The coolers *3* and *4*, made of beryllium bronze, have a good thermal contact with the shell of the high-pressure bomb *6* made of stainless steel. The stationary property of the regime is maintained by liquid *7* which circulates on the outer surface of the bomb at different temperatures. The temperature gradient on samples was studied by the copper-constantan thermocouples T_{I}, T_{II}, T_{III} and T_{IV}, which are inserted into the boring *8* in the cooler and the case of the furnace at the distance of 0.2 mm from the butt-end of the samples. The boring *8* were stopped by epoxide resin plugs to eliminate the possible effect of pressure on the readings of the thermopairs. In order to improve the thermal contact, the butt-ends of the samples, coolers and furnace were thoroughly polished and pressed to one another by the curragated tube *9* with apertures for liquid. The thermal losses from the side surfaces of the gradient heater *2* were compensated by the furnace *10* on the insulator *11* by means of the differential thermocouple. The pressure, created by the liquid in the bomb, was measured by the manganine manometer *12*. Castor oil *5* was used as hydrostatic liquid, which was pumped into the bomb through the thick-walled tube of stainless steel with carbine *13*. The carbine was pressed to the shell of the bomb through paddings *14* by condensing stopper *15*. All wires left the system through aperture *16* in the case of the cooler *4* and condensing stopper *17*. The boring in cooler *16* was scaled by epoxide resin. To measure heat loss from the side surfaces of the samples, the thermal conductivity of castor oil was studied at different temperatures and pressures by means of Amirkhanov's absolute stationary method [1]. The maximal error of the experiment calculated taking into account the loss through castor oil, did not exceed 4%.

The paper presents and discusses experimental data regarding the influence of hydro-

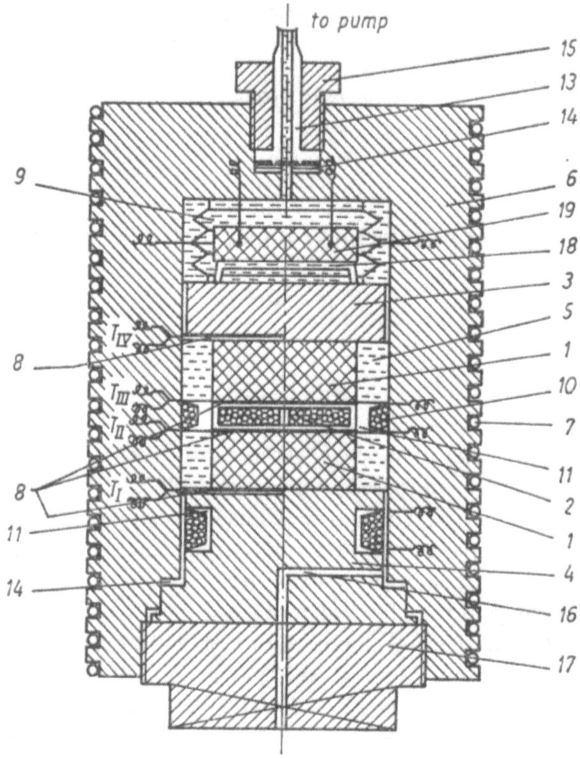

Fig. 1. The scheme of experimental device.

static pressure and temperature on the thermal conductivity of samples of amphibolite and andesite by the method described earlier.

Fig. 2 shows the dependence of thermal conductivity of amphibolite samples on pressure at different temperatures. The pressure causes nonlinear growth of thermal conductivity by 11 ÷ 13%. The obtained dependence of thermal conductivity on pressure tends to saturation. The growth of temperature causes reduction of thermal conductivity.

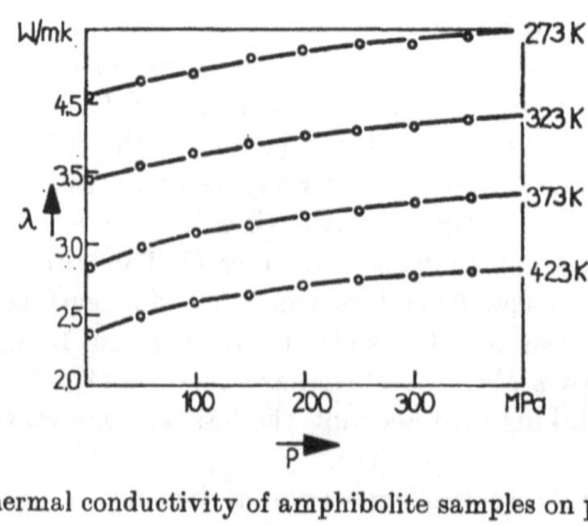

Fig. 2. Dependence of thermal conductivity of amphibolite samples on pressure and temperature-

Fig. 3 presents the dependence of thermal conductivity of amphibolite on temperature at atmospheric pressure and at 3.5 kbar. Under atmospheric pressure it has the character of inversely proportional dependence and obeys the Eiken's law ($\lambda \approx T^{-n}$, where $n = 1$). Therefore, the heat transfer in the studied samples is carried out mainly by means of lattice conductivity.

With the growth of hydrostatic pressure the character of temperature dependence of thermal conductivity also changes. As seen in Fig. 3, hydrostatic pressure causes a certain fall in temperature dependence of thermal conductivity and deviation from the Eiken's law. Therefore, if at atmospheric pressure the exponent of power $n = 1$, then at 350 MPa $n = 0.8$.

The obtained experimental data show, that the regularity in the distribution of heat in the amphibolite samples change under pressure in the same way as in semi-conductors and dielectrics [2, 3, 4, 5].

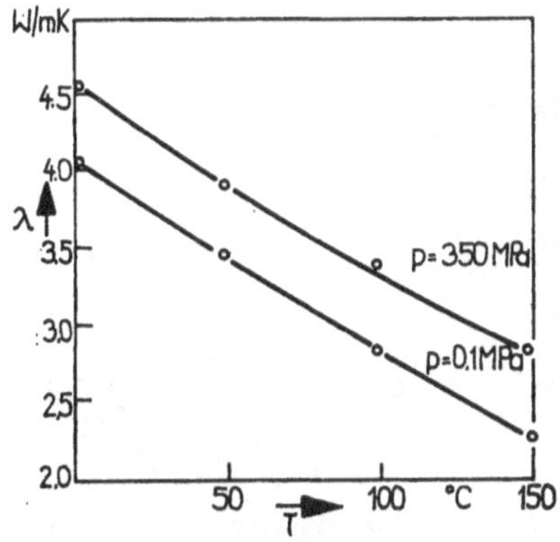

Fig. 3. Dependence of thermal conductivity of amphibolite samples on temperature.

Since the value of thermal conductivity directly depends on the rate of propagation of longitudinal ultrasonic waves, the observed change can be attributed to the change in the velocity of ultrasound in samples. At the present time, in the literature a large amount of experimental material is being published dealing with the effect of pressure on the rate of propagation of longitudinal ultrasonic waves in rocks [6]. It was noted, that on the curves of dependence of the ultrasound velocity on pressure two intervals can be distinguished and that the transition from one interval to another corresponds to the pressure 1.5 $\div$ 2 kbar, or to the depth of bedding 4—6 km [6, 8].

The studies described in paper [7], show that hydrostatic pressure increases the mobility of crystals dislocations and causes their displacement to the boundaries of crystallites. "The dislocation wall" formed on the boundary of crystallites in the polycrystalline substance, as the result of compression, can be one of the causes of non-linear dependence of thermal conductivity on pressure and of deviations from the Eiken's law.

It is interesting to study the dependence of thermal conductivity on hydrostatic compression in the substance, which according to data of other authors [6] is inhomogeneous in structure and contains about 80% of matter in glassy state.

Experimental data on the dependence of thermal conductivity of andesitic samples on temperature show that the rise of temperature to 423 K causes a slight increase in thermal conductivity (by 2–3%). The low dependence of thermal conductivity on temperature implies the presence of glassy substance in the studied andesitic samples.

Experimental data on the dependence of thermal conductivity of andesitic rock samples on pressure are given in Fig. 4. This Figure shows that the hydrostatic pressure increases thermal conductivity by 6 ÷ 7%, and the curve of dependence of thermal conductivity on pressure can be divided into two straight intervals, where the gradual transition from one interval to the other corresponds to the 180 MPa pressure.

A visible decrease of baric dependence of thermal conductivity in andesitic samples at $P = 180$ MPa indicates the appearance of the phenomena causing obvious dispersion of heat waves during these compressions.

The dependence of thermal conductivity on pressure, obtained by us, can be determined as the influence of pressure on the rate of propagation of longitudinal ultrasonic waves in the andesite samples; this dependence is in reasonable concordance with the results obtained in paper [8].

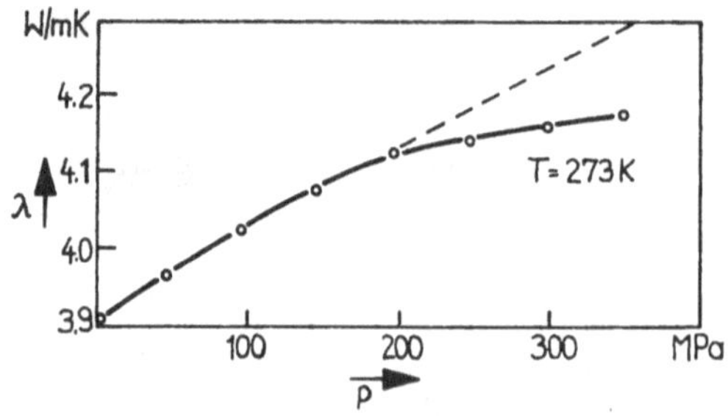

Fig. 4. Dependence of thermal conductivity of andesite samples on pressure.

References

[1] SADOVSKIY, I. A.; SARDAROV, S. S.: Co-subordination and similarity of geomovements in relation to the natural divisibility of rocks (in Russian). – DAN SSR **250** (1980) 846–848.

[2] AMIRKHANOV, K. I.; MAGOMEDOV, YA. B.; EMIROV, S. N.: Effect of allround pressure on the thermal conductivity of the tellur (in Russian). F.T.T. (1974) 5, 1512–1515.

[3] AMIRKHANOV, H. I.; MAGOMEDOV, I. B.; DAUNOV, M. I.; EMIROV, S. N.: The effect of hydrostatic pressure on some kinetic properties of $CdSnAs_2$. – High Temperatures – High Pressures 7 (1975) 6, 690–691.

[4] AMIRKHANOV, H. I.; MAGOMEDOV, YA. B.; EMIROV, S. N.: The change of thermal conductivity of $Cd_{0.28}Hg_{0.72}Te$ under hydrostatic pressure (in Russian). F.T.T. **17** (1975) 2938–2940.

[5] AMIRKHANOV, H. I.; EMIROV, S. N.; KRAMYNINA, N. A.: The study of the dependence of thermal conductivity of antimonide of gallium on pressure (in Russian). F.T.T. **21** (1983) 2486–2488.

[6] The physico-mechanical properties of rocks and minerals under high pressures and temperatures (in Russian). Nauka (1974).

[7] STRELTSOV, V. A.: Hydrostatic pressure effect on dislocation (in Russian). Physics and technology of high pressures (1982) 11, 21–24.

[8] The physical properties of rocks and minerals under high pressures and temperatures (in Russian). Nauka (1978).

Physical mechanisms affecting the temperature dependence of thermal transport properties of rock samples

U. SEIPOLD

The combined investigations of the temperature and pressure dependence of the thermal transport properties have shown that within the Earth's crust the first influence plays the overwhelming role. For this reason we looked at the different mechanisms causing the temperature dependence. If we define a rock as a heterogeneous dielectric we can neglect the effects of conducting electrons and defect electrons. We also assume that there are no important components with magnetic order. Under these circumstances the transport of thermal energy takes place by phonons and possibly photons and excitons. At first we concentrate the attention on the contribution of the phonons. If there is no interaction between the phonons and no interactions between phonons and lattice imperfections in the most common sense then the thermal resistivity is zero. In order to explain the finite value of the transport properties one has to look at the different scattering processes of the phonons. The effect of the different processes depends on the wavelength and for this and other reasons on the temperature. The interaction between the phonons is caused by the anharmonicity of the crystal lattice that means by terms in the lattice potential higher than of the second order. The third order term gives interactions of 3 phonons. Here we have to differentiate between normal and Umklapp-processes, where only the first ones are momentum conserving. According to Klemens the Umklapp-processes are most important for the transport of thermal energy. Leibfried and Schloemann have shown that for monoatomic crystals of cubic symmetry the T^{-1}-law is valid for the temperature dependence. For more complex substances one has to take into account the effects of the much more complicated phonon spectra, especially the occurence of optic phonons. For silicates as well the structure is very different from the cubic symmetry. According to the experimental experiences more complex substances have a lower temperature dependence. The L.-S.-law has to be generalized to

$$\lambda \propto T^{-n} \quad \text{with} \quad n < 1. \tag{1}$$

On the other hand, for compounds with ionic bonding and simple cubic structure with a small number of atoms in the unit cell we often have the case $n > 1$. For example, we found $n = 1.14$ for potassium bromide in the NaCl-phase, $n = 1.28$ in the CsCl-phase. It was possible to explain these values by the so-called volume effect. The theory is developed under the condition of constant volume. In reality the measurements are carried out at constant pressure. So we have to take into account thermal expansion and corresponding changes of lattice forces. Our estimations of this effect have shown that it can cause values of the exponent n up to 1.35 for *KBr*.

Terms of higher order in the lattice potential lead to phonon interactions of higher order, for example 4-phonon-processes. They give a temperature dependence proportional to T^{-2}. But we believe that for rocks the interactions of phonons with lattice imperfections including grain boundary effects should have a particular meaning. In silicate minerals we have high concentrations of point defects of different types. Every defect works as a scattering center for the phonons. The solid state theory gives results for the extreme cases of low and high defect concentrations. In petrophysics the latter case should be realized. Then the thermal conductivity is given by

$$\lambda \propto (cT)^{-1/2}, \tag{2}$$

c = concentration of point defects.

This means an essentially reduced temperature dependence compared to the 3-phonon-scattering. If we look at our results concerning the temperature dependence we can confirm this prediction. In most cases the exponent n is much smaller than 1 (Table 1). All measurements were performed at high pressure to prevent the occurence of cracks by anisotropic thermal expansion.

Two-dimensional lattice imperfections dominate the transport properties at very low temperatures, if the phonon wavelength is in the order of the geometric dimensions of the defects. In this temperature range the 3-phonon-processes have vanishing influence. In the high temperature approximation the temperature dependence resulting from boundary scattering is low because of the constant free path length. It should depend on the type and the special dimensions of the imperfections.

Many discussions are concerned with the role of radiative heat transfer. For this case from the very simplificated theory follows for the corresponding contribution to the thermal conductivity

$$\lambda = \frac{16n^2S}{3\bar{\varepsilon}} T^3, \tag{3}$$

n = refraction index,
S = Stefan-Boltzmann-constant.

$\bar{\varepsilon}$ is the mean value of the extinction coefficient over the wavelength. $\bar{\varepsilon}$ is composed of two parts:

$\bar{\varepsilon} = \alpha + \sigma$,
α = absorption,
σ = scattering coefficient.

Equation (3) shows that the completely different temperature dependence should allow to detect this part. This possibility was tested by us for potassium bromide. *KBr* has a high transparency in a wide infrared range of the spectrum. In fact an additional T^3-dependent part was detected at high temperatures (Fig. 1). In contradiction, in most cases for rock samples we found a monotonic decrease of the conductivity with increasing temperature up to about 1000 K. The results of other authors agree with this observation. Theoretical estimations of the radiative heat transfer take into account mostly only the spectral dependence of the absorption measured for mineral samples of

Table 1. Exponents n and m for the temperature dependence of the thermal conductivity and diffusivity respectively

Sample	p (GPa)	n	m
Granite	0.075	0.32	0.69
Kirchberg	0.3	0.39	0.73
	0.75	0.45	0.81
Granite	0.1	0.53	0.91
Ruhla	0.3	0.13	0.49
	0.75	0.29	0.63
	1.1	0.25	0.58
Plagiogranite	0.1	0.61	0.98
	0.3	0.45	0.81
	0.75	0.45	0.78
	1.1	0.37	0.71
Granodiorite	0.075	0.675	1.04
	0.3	0.90	1.24
	0.75	0.94	1.28
Quartzite	0.05	1.33	1.74
	0.3	1.23	1.56
	0.75	1.05	1.40
Plagioclase	0.3	0.19	0.54
	0.75	0.18	0.81
Granulite	0.125	0.355	0.77
Höckendorf	0.3	0.59	0.97
	0.75	0.68	1.07
Pyroxene	0.3	0.38	0.77
Granulite	0.75	0.49	0.87
	1.5	0.44	0.83
Granulite	0.125	0.26	0.61
Röhrsdorf	0.375	0.20	0.56
	0.75	0.29	0.64
Granulite	0.05	0.61	
Röhrsdorf	0.3	0.705	
	0.75	0.80	
	1.5	0.68	
Granulite W	0.1	0.24	0.65
	0.3	0.205	0.60
	0.75	0.15	0.50
	1.75	0.125	0.45

Sample characterisation

Granite Kirchberg: Quartz 38, potassium feldspar 31, plagioclase 29, biotite 1, muscovite 1 per cent. Fine grained monzogranite with not orientated panxenomorphic granular fabric.

Granite Ruhla: Quartz 23, plagioclase 41, potassium feldspar 35, biotite 1 per cent. Medium grained monzogranite with not orientated panxenomorphic granular fabric.

Table 1. (Continuation)

Granodiorite: Quartz 17, plagioclase 47, potassium feldspar 23, biotite 13 per cent. Medium grained biotite granodiorite with not orientated hypidiomorphic granular fabric.

Granulite Höckendorf: Quartz 40, plagioclase 33, perthite 17, garnet 8, biotite 1 per cent. Fine grained granulite with clear parallel structure.

Pyroxene granulite: Pyroxene 50, plagioclase 40, garnet 3 per cent.

Granulite Röhrsdorf: Quartz 47, plagioclase 23, perthite 16, garnet 3 per cent.

Granulite W: Quartz 38, plagioclase 9, potassium feldspar 47, biotite 1, garnet 4 per cent. Fine grained granulite with clear parallel structure.

high crystal perfection (gemstone quality). But it has to be considered that in rock samples the extinction is much higher because of the high contribution caused by scattering. This influence is important especially for fine grained rocks. We believe that the contributions the excitons pay to the thermal conductivity is negligible for the temperatures existing in the crust, because the concentration of electrons and defect electrons is too small.

If we want to compare the temperature dependence of thermal conductivity λ and diffusivity K we have to consider that between both values there exists the correlation

$$\lambda = c\varrho K, \tag{4}$$

ϱ = density

(in strong sense valid only for isotropic homogeneous solids). Estimating the temperature influence on the different members in this equation we found that only the specific heat c has to be corrected. When comparing the results in literature we derived the approximation formula

$$c(T) = c_0 f(T).$$

c_0 is a constant characterizing the special mineral or rock, $f(T)$ is a universal function.

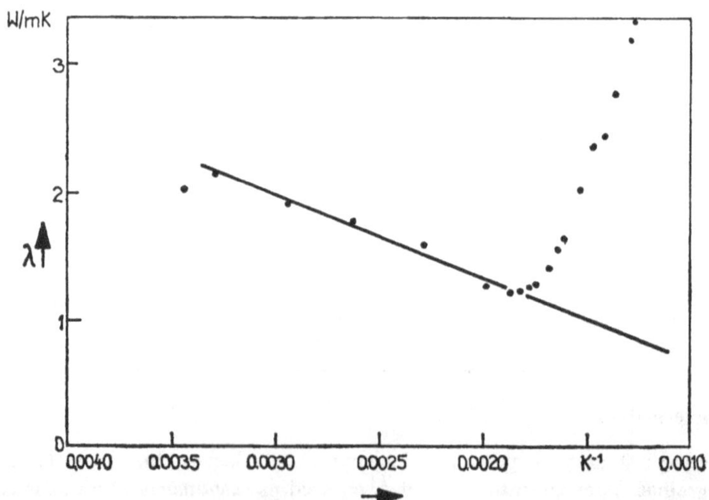

Fig. 1. Temperature dependence of thermal conductivity for *KBr*

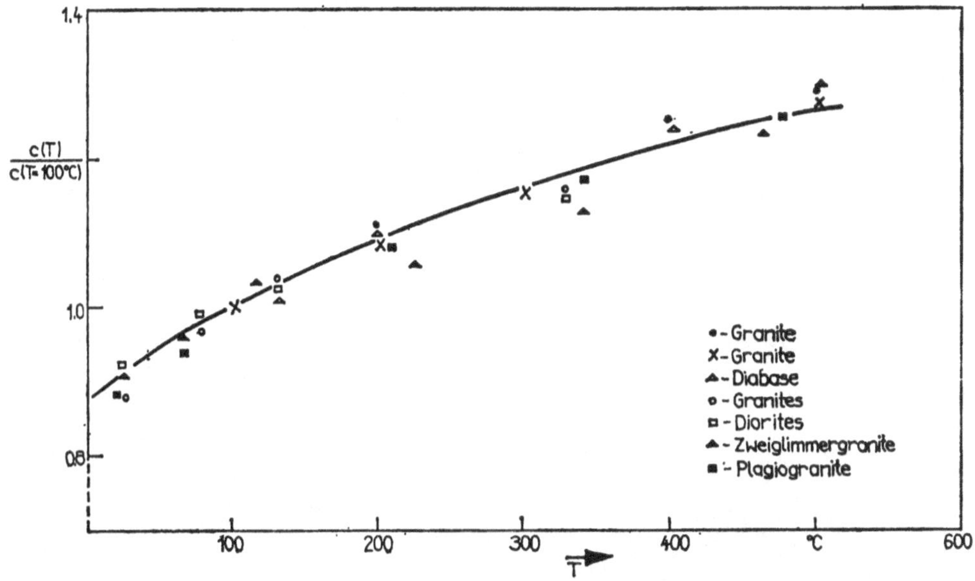

Fig. 2. $f(T)$ for rocks of the upper crust

$f(T)$ is an increasing function with growing temperature (Fig. 2). So it can be developed in a power series of the temperature with positive exponents. From this it follows that the exponent n in $\lambda \propto T^{-n} \propto T^{+x} \cdot T^{-m}$ is smaller than the exponent m in $K \propto T^{-m}$. This means that the temperature dependence of the diffusivity is always higher than the corresponding variation of the conductivity.

References

LEIBFRIED, G.; SCHLOEMANN: Wärmeleitung in elektrisch isolierenden Kristallen. Nachr. Akad. Wiss. Göttingen, Mathem.-Phys. Klasse IIa (1954), 71–93.

KLEMENS, P. G.: Theory of the thermal conductivity of solids. In: TYE, R. P.: Thermal conductivity. Academic Press London, New York 1969, Vol. 1, 2–68.

CHAPTER IV

Magnetic Properties

Change in mineral matter remanent magnetization at high *P* and *T*

T. S. Lebedev, B. Y. Savenko

The study of stability of natural remanent magnetization (I_n) of rocks to tension field effects is of great importance in the solutions of some problems of magnetism and paleomagnetism.

Several sets of experimental studies have been made to elucidate effects of stressed states of rocks on the I_n and remanent saturation magnetization [I_{rs}] changes at uniaxial compression [1, 2] under high hydrostatic pressure (P) [2, 3]. Maximum pressure (250 MPa) used in our experiments did not exceed the elastic deformation range of the samples studied. The experimental methods and technique are described in [4].

Fig. 1 illustrates results of experimental high hydrostatic pressure effects on changes of natural remanent magnetization of a number of rocks. A different I_n decrease has been established for samples of different petrographic composition. At 250 MPa effusive rocks retain 0.4—0.9 and ferruginous quartzites 0.2—0.8 of their initial I_n values. For some rocks, the pressure of the order of 100 MPa was sufficient for I_n to become zero. The first thing required to find causes of different $I_n = f(P)$ behaviour is studying

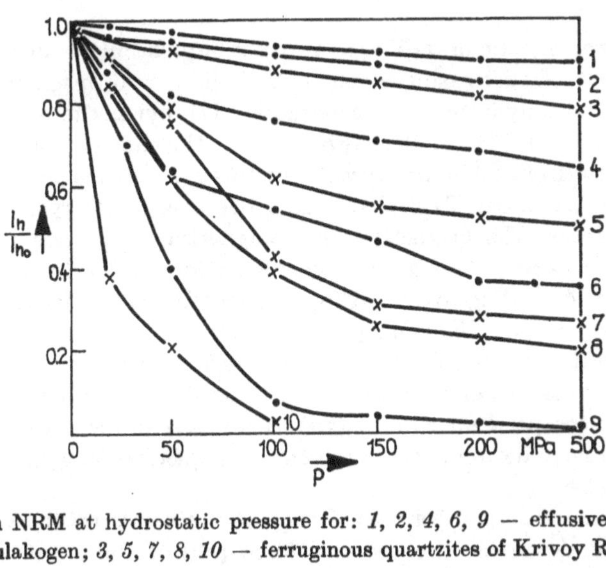

Fig. 1. Change in NRM at hydrostatic pressure for: *1, 2, 4, 6, 9* — effusives of Dnepr-Donetz aulakogen; *3, 5, 7, 8, 10* — ferruginous quartzites of Krivoy Rog.

physical processes creating the magnetic state of a rock, hence degree and character of change in its remanent magnetization.

For granites [4] direct dependence of the degree of destruction of their laboratory remanent magnetization at the simultaneous *P-T* effect on its type has been established. Most, intermediately and least stable were samples with TRM (I_{rt}), partial and viscous magnetization respectively. The magnetic prehistory of each sample includes the total sum of external effects exerted on it during its whole life. In the I_n formation, *P* and *T* are of great importance. The higher *P* and *T* to which the rock was exposed during its formation, the greater the I_n stability to pressure in experiments. Such a relation between the thermobaric condition of rock formation and $I_n = f(P)$ behavior in the experiment has been revealed in studies of ferruginous quartzites of Krivoy Rog [5]. The lowest I_n stability was recorded for samples containing magnetic of low-temperature mineral associations the highest one was that of high temperature ones.

Mineralogic investigations of the magnetite of ferruginous rocks of different facies show their structure to change with increasing metamorphism. Magnetite forms in different *P-T*-situations creating different typomorphic features expressed in size, inner grainstructure, mechanic inclusions and isomorphic impurities. This is the main factor providing wide magnetic properties stability ranges for ferruginous quartzites at high *P* and *T*, the stability of their natural remanent magnetization (NRM) depending mainly on size, form and coordination of grains in the rock as well as on metamorphic transformation.

The RSM of the samples studied is also characterized by irreversible changes occuring when hydrostatic pressure is applied (Fig. 2) whose increase lowers I_{rs} values differently for petrographically different rocks. In *P* experiments, the NRM stability to *P* is originated by NRM formation process whereas I_{rs} observed in high magnetic fields was favoured by nearly similar energetic state of the domain structure of ferrimagnetic. Therefore I_{rs} stability to pressure will depend largely on the composition and structural features of magnetic minerals (e.g. magnetite and pyrrhotite) but first of all on density and order of internal defects and dislocations as well as their mobility and external stress effect.

The type of ferrimagnetic prevailing in a rock is of great importance in estimating I_{rs} stability to pressure I_{rs} decreasing by the hyperbolic law for magnetite-bearing rocks. The I_{rs} curves differ only in the curvature radius. The pyrrhotite-bearing rocks show a more complicate $I_{rs} = f(P)$ form. Such differences in the character of pressure-demagnetization of I_{rs} arise from different crystal structures of ferrimagnetics and associated internal stress distribution. Typical of magnetite are cubic crystals with magnetic anisotropy properties. The magnetization-demagnetization process occurs differently but over all crystal axes. The higher magnetostriction of this ferrimagnetic makes it pressure-sensitive. According to STACEY [6] $1 \cdot 10^{-5}$ MPa suffice for remanent magnetization decrease by 0.05%. A. KAPIČKA [7] defines pressure capable of changing magnetic features of magnetite-bearing rocks to be ca. 50—60 MPa. The otherwise shaped rhombic pyrrhotite crystals are rectangular prisms magnetized and consequently demagnetized only in one plane (perpendicular to its axis) irrespective of field direction.

An important role in destroying rock remanent magnetization by pressure plays grain size of ferrimagnetic.

The size and the structural features of magnetic minerals within a compositionally intermediate rock determine the coercive force (H_c) of the former. Like remanent

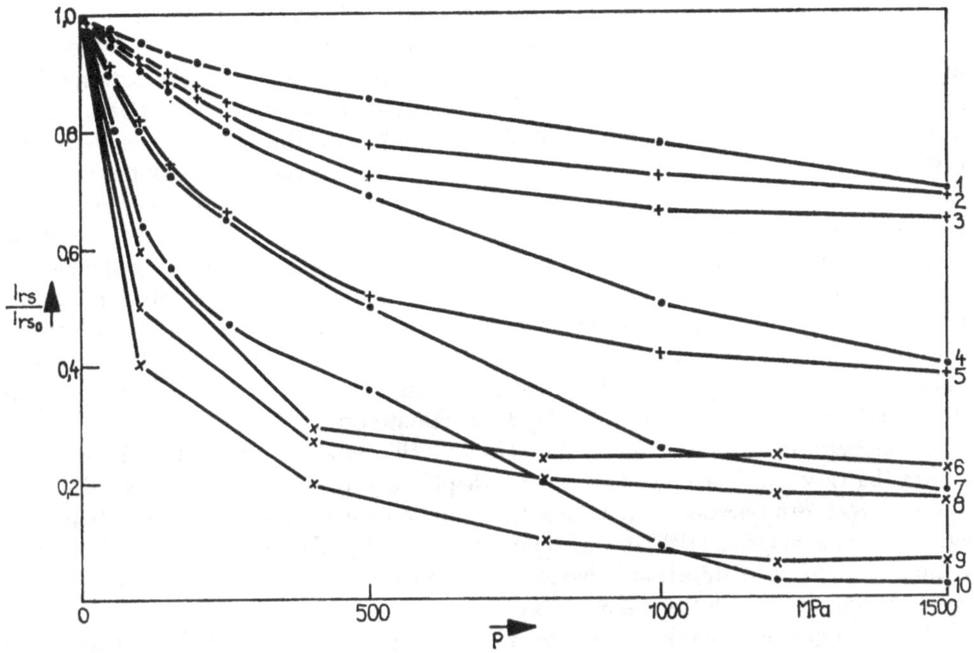

Fig. 2. Irreversible changes in RSM of some rock samples at hydrostatic pressure. *1, 4, 7, 10* — volcanogenic rocks of Transcarpathians; *2, 3, 5* — effusives of Dnepr-Donetz aulakogen; *6, 8, 9* — ferruginous quartzites of Krivoy Rog.

magnetization (RM) the H_c value is closely related with internal stresses causing energy equilibrium of the ferrimagnetic. Therefore its ratio of reversible to irreversible I_{rs} variation components is directly connected with magnetically soft and magnetically rigid domain distribution in the spectrum of microcoercive forces of the ferrimagnetic at high pressure. Frequently applied pressure eliminates the irreversible I_{rs} part and stabilizes internal stresses. A similar result can be obtained after compression of rock above its flow limit. The samples effected preliminarily by pressure response much more weakly to its subsequent effects to which their RM appears to be more stable.

The mechanism of RM recovery by frequent rock compression is associated with internal stresses and may be explained by magnetostatic interaction of domains with predominant crystal anisotropy. At uniaxial pressure, the domain structure reorientation is observed in the direction perpendicular to compression. The hydrostatic pressure effect removes internal stress. However, spontaneous magnetization vector orientation occurs under the orientating laboratory magnetic field effect. As domain-boundary movements break especially in dislocation accumulation zones, the latter show a different degree of recovery of the reversible component of RM changes.

Internal stresses originate from imperfect crystalline structure of ferrimagnetic mineral grains due mainly to dislocations, point defects and their accumulations. The effect of these factors on the degree of I_{rs} changes depends on character and type of dislocation distributions as well as on the domain-boundary category. The external length changes the ferrimagnetic dimensions according to the magnetostriction constant sign. Local changes of magnetization causing additional stresses also occur within the domain boundaries. Their interaction with dislocation tension field breaks the domain boundary movement and determines the coercive force value. When the external load reaches H_c values discrete dislocation of domain boundaries from one potential minimum to the other, hence the change of RM of the rock occur. The experiments show that these processes can be reversible and irreversible. The irreversible I_{rs} changes are associated with additional energy losses (socalled hysteresis losses). The reversible ones are mainly associated with 180° angle of domain boundaries at which the domains are not greatly affected by internal stress field. At the same time, the reversible changes are most greatly characterized by spontaneous magnetization vector rotation. Under laboratory magnetic field conditions, a ferrimagnetic includes boundaries of two types. The first-type boundaries occur when the resultant directed towards the increasing volume of in-field magnetized domains acts from the side of the external magnetic field, while the second-type ones display force oriented to decreasing volume of counter field magnetized domains. Under the effect of the external magnetic field equaling energy barriers produced by dislocations displaces the boundaries additionally. The RM of a rock can decrease or increase depending on the first-to-second type boundary ratio. A strong magnetic field in which initial I_{rs} is obtained orients all the ferrimagnetic domains in the field direction. Therefore in studying $I_{rs} = f(P)$ only remanent saturation magnetization (RMS) decrease is observed.

Unlike magnetic minerals, the rocks exhibit more complicate RM changes at elastic stress since the external compression acts on a ferrimagnetic via a non-magnetic matrix. Differences in elastic features of the main rock-forming minerals cause nonuniform deformation of the whole matrix when effected by pressure, which influences the change of the magnetic properties of the rock differently. This should be taken into account when magnetic parameters of rocks are correlated at high pressure as well as in comparing the effect of different demagnetizing factors (e.g. alternating magnetic field and pressure) on RM. The $I_{rs} = f(P)$ stability increase with growing ferrimagnetic force may only be expected for samples with almost identical elastic features of nonmagnetic matrix minerals. If alternating magnetic field acts immediately on I_{rs} of a ferrimagnetic the pressure effect greatly depends on the rock-forming mineral features at the external load transfer. Therefore the effect of ac demagnetization of RM of a sample is higher than that of external compression.

The changes in mineral substance remanent magnetization depend on physico-chemical features and are also caused by some other factors of which the main are technical conditions of experiments. Of great importance is the mode of external pressure transfer. The experimental results [1] show higher $I_{rs} = f(P)$ gradients to appear with applying uniaxial pressure. All the samples studied show the following characteristic features. The highest I_{rs} values are observed at slightly elevating P but they become moderate with further P increase to maximum values used in the experiment. Another picture arises with hydrostatic pressure effects: initially I_{rs} changes are insignificant, the most intensive ones occurring within the relatively high P range. The simultaneous continuous hydrostatic and uniaxial pressure effect places the magnetization curves between the extreme magnetization values [7]. Note also that irreversible changes caused by uniaxial compression are more significant than the total I_{rs} changes occurring at hydrostatic pressure.

Irreversible I_{rs} changes whose growth is greatly defined by pressure intensity play crucial role in rock destruction. They are most intensive in hydrostatics at 500 MPa. The further pressure increase does not cause significant magnetization changes.

The contribution of reversible processes to the total rock demagnetization effect also depends on the mode of pressure application. The hydrostatic pressure up to 200 MPa used in our experiments produced the highest reversible component reaching 14% of the total I_{rs} (its mean value being 6%). The reversible process contribution is less at uniaxial compression extending the maximum I_{rs} changes to 38% and the average ones to 24%.

The experimental results revealed one more feature of the RSM behaviour at pressure, i.e. changes in I_{rs} stability at preliminary thermal processing of samples reorientating low-coercivity domains, redistributing energy barriers and displacing domain boundaries in microcoercive force spectrum. The experimental results show that only the application of pressure alone stabilizes the internal stresses of a ferrimagnetic in most samples (curves $I_{rs} = f(P)$ approaching the asymptote) which, however, is insufficient for complete destruction of their magnetization.

Pressure as demagnetizating factor has been estimated by experiments with magnetite-titanomagnetite bearing rock samples effected by uniaxial pressure (to 400 MPa) and temperature up to 600 °C [8]. Ferruginous-quartzite samples were investigated. Their main ferrimagnetic is magnetite with Curie-temperature of ca. 580 °C. The samples were effected by uniaxial pressure to 50 MPa at different temperatures. The I_{rs} changes were recorded after each 100 °C interval. In the first 100 °C interval magnetization is more significantly destroyed by pressure. Heating to 200 °C yields nearly equal effect of each demagnetizing factor. Above 300 °C, only temperature alone decreases the rock magnetization. Samples preliminarily heated to 200 °C and 300 °C lose their magnetization completely at 170 and 90 MPa respectively. Nearly the same features are characteristic of titanomagnetite-bearing rocks with Curie-temperature of 410 °C. The application of pressure of the order of 50 MPa eliminates ca. 15% of the initial magnetization. Heating to 100 °C (without P application) decreases the magnetization by almost 20%. At 200 °C the temperature effect in I_{rs} destruction increases notably; the magnetization changes by 5 and 20% respectively. Above 300 °C the rock magnetization is mainly decreased by elevating temperature.

Thus, the experimental results show the I_{rs} changes to be most significantly affected by pressure at low temperature, P-contribution decreasing with T-increase and 300 °C being probably the lower limit of pressure influence on the RM of rocks.

On the basis of experimental data [5, 9—13] a comparative analysis of changes in RMS of crystalline rocks of some geologic regions of the Ukraine with depth has been made. The data of table illustrate the degree of I_{rs} preservation for rock samples to depths of the order of 10 km for which [14] shows P and T values of all (except for Transcarpathian) Ukraine regions to be 200—250 MPa and 150—200 °C respectively. For Transcarpathians temperature reaches 300 °C at the afore mentioned pressures. The highest scatter of P values does not exceed 50 MPa for the regions studied.

An analysis of the data of Table 1 shows all the rocks studied to decrease their RM with depth. At a depth of 10 km a rock (unaffected by temperature) can preserve 47—60% of its initial magnetization. The behaviour and changes of the I_{rs} value with depth also depend greatly on the composition, structural and genetic features of ferrimagnetic minerals.

The main factors determining the I_{rs} stability to pressure, hence I_{rs} distribution with

Table 1. Changes of remanent saturation magnetization (I_{rs}) in % for some rocks of the Ukraine. (The numerators show the minimum and maximum and the denominators the average $I_{rs}(H)$ changes (in %) as determined by extreme values).

Rocks, Regions	Depth (km)				
	2	4	6	8	10
Crystalline rocks of the Dnepr-Donetz aulakogen basement	95—70 / 82	95—60 / 78	90—50 / 70	85—45 / 65	80—35 / 57
Effusives of Dnepr-Donetz aulakogen	90—65 / 82	90—50 / 70	85—45 / 65	80—35 / 57	75—25 / 50
Intrusive rocks of the Ukrainian Shield	95—75 / 85	95—65 / 80	95—45 / 70	95—30 / 62	95—15 / 55
Effusives of the Ukrainian Shield	95—80 / 87	95—40 / 67	90—35 / 62	85—30 / 57	85—25 / 55
Metamorphic rocks of Krivoy Rog	95—80 / 87	90—50 / 70	80—40 / 60	70—30 / 50	65—30 / 47
Intrusive rocks of Crimean highland	95—70 / 82	85—55 / 65	80—45 / 62	80—40 / 60	75—40 / 57
Volcanogenic rocks of Transcarpathians	98—75 / 88	95—60 / 77	90—50 / 70	90—40 / 65	90—30 / 60

depth for magnetite-bearing rocks of the Ukrainian Shield and metamorphic rocks of the Krivoy Rog region are the thermodynamic conditions of their formation. Magnetite forms under different *P-T* conditions expressed in its different typomorphic features, i.e. in size, internal grain structure inclusions and isomorphic impurities, which expands the range of stability of their magnetization to pressure.

The study of the features of I_{rs} changes in quartzites of different metamorphism facies at high pressure has provided additional information on *P-T* effects possibly exerted during their formation and evolution [5]. The experiments made at relatively low pressure show the lowest I_{rs} stability to be that of quartzites whose magnetite associates with high-temperature minerals sharply decreasing their magnetization with pressure increase to 100 MPa. Higher pressures lower the I_{rs} decrease gradients, the curves approach the asymptote, which means RM value stabilization. Such RM changes are probably caused by prolonged paleoeffects of high *P-T*. The sharp I_{rs} decrease occurring in the initial range (to 100 MPa) can also be associated with particular formation of magnetite whose grains are almost completely void of visible inclusions.

RSM of mean-temperature ferruginous quartzites appeared to be more stable to pressure probably due to low-*P* and *T* effects during their formation under *P-T* conditions characteristic of amphibolite facies. The increase of I_{rs} stability to high pressure is also favoured by oriented amphibole and biotite inclusions of magnetite.

The RSM of ferruginous quartzites of the green-schist facies appears to be the most stable to high pressure especially in the initial interval (to 100 MPa).

Further pressure elevation slightly decreases the I_{rs} probably due to non-metallic mineral inclusions existing in magnetite grains. In studying high-pressure RSM, changes of nearly identical magnetic state of magnetite domain structure of different ferruginous quartzites were considered. Thus, the experimental data may be used as additional information on the composition, structural and genetic features of magnetite.

Known P and T distribution in deep zones of crustal areas, changes in magnetite-bearing rocks with depth can be more completely characterized by experimental and theoretical RM data. At hydrostatic pressure of 500—800 MPa (depths of the order of 20—30 km) such rocks of the Ukrainian Shield possess on the average 20—30% of their magnetization.

The $I_{rs} = f(P)$ stability also depends on the effect of secondarely imposed processes during the geological "life" of a rock. The influence of this factor was inferred from studies of intrusive rocks of the Crimean highland [12]. Within the tectonic structures including intrusive bodies allometasomatic processes of natrium branch are highly developed whose intensity is different in different areas. That is why the rocks are differently transformed metasomatically. Epigenetic isolations of fine-dispersed magnetite formed by natrium metasomatism in paramagnetic minerals and which exist together with syngenetic titanomagnetite crystals and exert the predominant effect on the magnetic properties of a rock are also noted. The degree of the metasomatic transformation of the rock is expressed in the stability of its magnetization to pressure. This enables us to use the data available as additional information on the composition, structural and genetic features of magnetite-bearing rocks.

The rocks containing minerals of the titanomagnetite series differ in a wide range of pressure-induced I_{rs} changes. Of all ferrimagnetics these minerals are the most complicate objects of high-pressure studies mainly because of their varied composition and textural features. This impedes analysis of all factors influencing internal stress distribution and determining the degree of $I_{rs} = f(P)$ changes of titanomagnetites. We report only a few of our experimental results.

For the Ukrainian Shield basalts as geologically, petrographically and chemically typical representatives of trap formation a clear relation between the character of RM change at high pressure and Curie temperature [11] is found. The magnetic minerals of basalts are represented by homogeneous titanomagnetites least affected by oxidation. Their stability to pressure increases almost linearly with decreasing Curie temperature, this increase depending on decrease of magnetite content in the $x\ TiFe_2O_4(1-x)\ Fe_3O_4$ system that increases magnetic mineral inhomogeneities when oxidation degree and other factors are equal. This greatly affects the magnetite grain size and form, grain position and, consequently, internal stress distribution favouring generally the stability to pressure.

A quite other picture is observed for Dnepr-Donetz alaukogen effusives belonging to hypabyssal facies rocks [10, 11]. Although titanomagnetite disintegration and oxidation yielded identical minerals the nature and conditions of these processes were different. Hence, the reaction products can contribute in different ways to entire magnetization effect in each concrete case. This makes it impossible to establish unambiguously the Curie-temperature dependence of the RM of a rock at pressure.

These particular magnetic features of effusive rocks are probably due to the degree of preservation of primary ferrimagnetic material at the time of investigations. Though the samples contain the same mineral (titanomagnetite) its formation condition and degree of secondary process evolution can cause magnetic parameter variation at pressure.

The stability of $I_{rs} = f(P)$ of volcanogenic rocks of Transcarpathians is mainly defined by grain size and structural features of titanomagnetite [13]. According to [15] volcanites contain sets of ferromagnetic grains whose sizes differ by several orders. Fine grains have typically pseudo-single-domain and single-domain structures. For magnetite the transition from single- to multidomain composition is near to the upper limit of the supermagnetic one. Moreover, the multidomain structure of primary homogeneous titanomagnetites can be replaced by the single- and pseudo-domain one when coarse grains fractionize by high-temperature oxidation.

The titanomagnetite grains and disintegration products are presented as an collection of 2—3 to 50 μ sized impregnants. Such size provides a high coercivity of the rock hence its high stability to pressure. In general, the titanomagnetite-bearing rocks appeared to be more stable to pressure and preserve, on the average, 42% of their initial I_{rs}.

Thus, several factors determining the degree of changes in RM of rocks effected by high pressure have been found and described. On this basis a close relation of their RM with genesis features has been established. Further minerals matter investigations aimed at studying magnetic properties at simultaneous *P-T* effect will evidently enable us to elucidate more completely the thermodynamic conditions of rock formation and state in deep crustal zones.

Summary

High *P-T* changes in RM of rock samples are analysed. Some factors determining the character and degree of these changes are established and distribution of RM of some crystalline rocks of several geologic regions of the Ukraine based on laboratory experimental data is considered. The data obtained evidence a close relation of magnetization changes with genetic features of rocks.

References

[1] KAPIČKA, A.; SAVENKO, B. Y.: Comparative characteristic of changes of RM of basalts at hydrostatic and uniaxial pressures (in Russian). Geoph. Zhurn. 1 (1979) 3, 57—61.

[2] PODROUŽKOVA, Z.; KROPÁČEK, V.; ŠTERBA, O.: Instrument for measuring remanent magnetic polarization of rocks under high hydrostatic pressures of up to 200 MPa. Geofysikalni sbornik XXIV (1976) 463, Akademia, Prague (1979) 281—288.

[3] LEBEDEV, T. S.; SAVENKO, B. Y.; SHAPOVAL, V. I.; KORCHIN, V. A.: An equipment for studying magnetic properties of rocks at hydrostatic pressure to 2,5 kb (in Russian). Geofiz. sb. Ac. Sci. Ukr. SSR, Kiev (1972) 46, 85—87.

[4] Investigation of Earth's substance physical properties at high thermodynamic parameters (in Russian). Naukova Dumka, Kiev (1977) 220 p.

[5] LEBEDEV, T. S.; SAVENKO, B. Y.; POZNANSKAYA, N. F.; STEBNOVSKAYA, Y. M.: Magnetic properties of ferrugineous quartzites of the Ukrainian Shield at high *P-T* (in Russian). Geofiz. sb. Ac. Sci. Ukr. SSR (1977) 79, 34—44.

[6] STACEY, F.: Earth's Physics. Mir, M. (1972) 342 p.

[7] KAPIČKA, A.: Faktory ovlivňujici stabilitu remanentni magneticko polarizace hornin při působeni vnějšich tlaků. Sbornik referatů 7. Colostatni konference geofyziků, Gottwaldov (1980) 61—65.

[8] YELENSKAYA, M. B.; SAVENKO, B. Y.: Experimental studies of RM destruction at simultaneous high P-T effect (in Russian). In: Physical properties of rocks at high thermodynamic parameters, Baku, Elm (1978) 218—219.

[9] SAVENKO, B. Y.: Changes in magnetic properties of rocks at high P and T (in Russian). In: Anomalies of nonmagnetic field and deep crustal structure. Transact. Int. Work. meeting. Naukova Dumka, Kiev (1981) 10—15.

[10] LEBEDEV, T. S.; SAVENKO, B. Y.; INDUTNY, V. F.: Character of changes in RM of rocks of Dnepr-Donetz depression basement at high hydrostatic pressures (in Russian). Geofiz. sb. Ac. Sci. Ukr. SSR (1974) 61, 89—96.

[11] SAVENKO, B. Y.; POZNANSKAYA, N. F.; KARZANOVA, A. Y.: Studies of effusives of Ukrainian Shield and Dnepr-Donetz depression at high P and T (in Russian). In: Paleomagnetism, magnetism, magnetic field. Kiev, Naukova Dumka (1976) 61—68.

[12] LEBEDEV, T. S.; SAVENKO, B. Y.; POZNANSKAYA, N. F.: Magnetic properties of some rocks at high hydrostatic pressure (in Russian). In: Regional petrophysical characteristic of the soviet Baltic Shield. Apatites. Publ. off. Kola region section Ac. Sci. USSR (1976) 132—137.

[13] LEBEDEV, T. S.; SAVENKO, B. Y.; SHAPOVAL, V. I.; BURTNY, P. A.: Magnetic properties of volcanogenic rocks at different P-T conditions (in Russian). In: Investigations of geomagnetic field and paleomagnetism. Kiev, Naukova Dumka (1983) 156—163.

[14] LEBEDEV, T. S.; SHAPOVAL, V. I.; KORCHIN, V. A.: Programmed modelling of deep thermodynamic conditions in experimental studies of mineral substance elastic properties (in Russian). In: Geophysical studies of Earth's crust and mantle of Ukraine. Kiev, Naukova Dumka (1975) 25—30.

[15] GLEVASSKAYA, A. M.: Magnetic minerals and magnetism of volcanites (in Russian). Kiev, Naukova Dumka (1983) 208 p.

Magnetic and mechanical properties of porous rocks under external hydrostatic pressure

A. Kapička, Z. Podroužková

Summary

The changes of the remanent magnetic polarization of rock specimens with an effective porosity of up to 10% under hydrostatic pressures of up to 200 MPa are investigated. It is proved that the stability of the remanent magnetic polarization of specimens, whose pores are not filled with the pressure medium (zero pore pressure) is much smaller than if the pore pressure is non-zero. This result agrees with the concept of effective stresses in porous substances and qualitatively corresponds to the changes in the ultimate pressure strengths for samples with different pore pressures we observed.

Introduction

Experimental studies in changes of the magnetic parameters of rocks are conducted under various pressures [1—3]. Considerable attention is being devoted namely to hydrostatic pressure, because it models relatively well the actual pressure effects on rocks in the Earth's crust. The relevant results form the basis, e.g., of interpreting seismomagnetic effects and local magnetic anomalies.

Most rocks do not have a compact structure, but contain a certain amount of pores, cracks, etc. Pores between particles, generated in the process of sedimentation and consolidation, are characteristic especially for sedimentary rocks, inter- and intracrystal cracks are created in igneous and metamorphosed rocks as a result of the different thermal, elastic and plastic properties of various materials, or due to their anisotropy. If these rocks are saturated by a liquid medium, their mechanical and, probably, also their magnetic properties may change substantially. If we restrict ourselves just to the physical interaction of the liquid medium with the rock, we find that the effect of the pore pressure on the values of the ultimate strength and magnitude of permanent deformation of porous rocks was already proved experimentally some time ago (see, e.g., [4]). On the contrary, measurements of the changes of magnetic properties under hydrostatic pressure have only been carried out on rocks with very small porosity, or no attention was devoted to this effect. The purpose of this study is to investigate to what extent the saturation of a rock with a liquid medium, and consequently different pore pressure, is capable of affecting the changes in remanent magnetic polarization under hydrostatic pressure. The differences in the ultimate strengths under combined hydrostatic and uniaxial pressure between rocks whose pores were and were not filled with a liquid medium were measured simultaneously.

Experiment

The measurements were made with a sample of Sazava-type granodiorites whose variable porosity went up to 10%. It was found by X-ray that the ferrimagnetic fraction was formed by magnetite with a volume concentration of less than 1%. Some of the magnetic and mechanical parameters of the granodiorites being studied are in Table 1 (saturated remanent magnetic polarization J_{rs}, A.C. magnetic field H_D required to demagnetize to 0.5 J_{rs}, whose values provide an idea of the microcoercive force spectrum in the separate specimens, the ultimate pressure strength σ_P, and the effective porosity Φ).

Table 1. Magnetic (J_{rs}, H_D) and mechanical (σ_P) characteristics, and the effective porosity (Φ) of the granodiorites under study

Specimen	J_{rs} (T)	H_D (10^4 Am^{-1})	σ_P (MPa)	Φ (%)
1	$2.3 \cdot 10^{-5}$	0.8	161	1.7
12	$1.8 \cdot 10^{-4}$	1.5	152	3.5
14	$1.2 \cdot 10^{-5}$	—	110	1.9
15	$8.1 \cdot 10^{-5}$	2.8	49	3.9
17	$0.8 \cdot 10^{-5}$	—	240	5.6
19	$9.8 \cdot 10^{-5}$	0.6	136	9.8

Cylindrical specimens, diameter 8 mm, height 10 mm, were used for the pressure experiments. Their front surfaces were ground planeparallel. The sample as a whole contained variously modified specimens of each granodiorite: a) dried and uncovered, b) dried and covered, c) water-saturated after drying (uncovered). The specimens were dried at 80 °C for 60 hrs, and the covering of the samples was effected by a thin layer of epoxy-resin paint. The saturation was carried out under atmospheric pressure by submerging the specimens gradually in water, the degree of saturation being checked on a current basis by weighing. The mechanical properties under uniaxial pressure were measured with an Instron deformation instrument which recorded the loading force and deformation.

A special high-pressure chamber was used for measuring the deformation characteristics under combined hydrostatic pressure up to 250 MPa and differential pressure up to 700 MPa [5]. A high-pressure magnetometer, whose principle is founded on the rotation of the loaded specimen within pick-up coils, was used to measure the changes of the remanent magnetic polarization under hydrostatic pressure of up to 200 MPa. The induction system of the pick-up coils yields a continuous record of the pressure variations of the remanent magnetic polarization. A cylinder-piston-type hydrostatic chamber was used to study the process of gradually filling of the pores under very high pressures of up to 1.5 GPa. The effective porosity was determined by saturating the specimens with water in vacuum, using the relations

$$\Phi = V_P/V_c, \qquad V_P = (M' - M)/\varrho_{v'},$$

where V_c is the total volume of the specimens, V_P the volume of the interconnected pores, M' the mass of the saturated specimen, M the mass of the dry specimen, and

Table 2. Maximum degree of saturation (S_w) under normal and other hydrostatic pressures

Specimen	S_w (%)	S_w^{I} (%) 200 MPa	S_w^{II} (%) 200 MPa	S_w (%) 1.5 GPa
1	47	52	70	92
12	40	51	59	71
15	41	52	58	69
19	51	34	40	46

ϱ_v the density of water. The values of the effective porosity of the studied granodiorites are shown in Fig. 1. The maximum degree of saturation of the rock, $S_w = V_s/V_p$, where V_s is the volume of the filled pores, was studied under gradual water-saturation of the specimens and normal pressure, using the relation

$$S_w = (M_2 - M)/\Phi \varrho_K V_c,$$

where M_2 is the mass of the partly saturated specimen, M the mass of the dry specimen and ϱ_K the density of the liquid. The same relation was used to determine the changes in the degree of saturation under high hydrostatic pressures of 200 MPa and 1.5 GPa. The repeated application of the same pressure to the specimens is designated I and II. The pressure medium was pure mineral oil. The values obtained are shown in Table 2.

One can see that, under normal pressure, 40–50% of the volume of effective pores are saturated by the liquid, and that the degree of saturation is even larger when the specimens are under pressure. However, not even if very high pressures are applied is complete saturation of the pores achieved and, therefore, the pressure in the pores is mediated by a liquid-gas system.

Magnetic and mechanical properties under hydrostatic pressure

The effect of various pore pressures on the ultimate pressure strength was studied in the first place. The different pore pressures were achieved by using coates samples and water-saturated uncoated samples. The differential uniaxial pressure required to disintegrate the specimens was measured under various hydrostatic pressures. With a view to the limited differential pressure in the chamber (max. 700 MPa), lower hydrostatic pressures between 50 and 100 MPa were applied. The most important results of these measurements are summarized in Table 3.

Table 3. Ultimate strength σ_p under combined stress for various pore pressures

Specimen	Hydrostatic pressure (MPa)	Pore pressure (MPa)	σ_p (MPa)
14	100	100	290
14	100	0	480
17	50	50	260
17	50	0	380
19	50	50	200
19	50	0	350

The ultimate strength of granodiorites, which is relatively low under external atmospheric pressure (see Table 1), increase in the presence of external hydrostatic pressure. However, at the same time, one can see that there are very substantial differences in the specimens with different pore pressure under the same hydrostatic pressures, the ultimate strengths of the coated specimens being systematically higher.

These differences in the porous materials can be explained quantitatively by introducing the so-called effective stress. If we consider a dry, homogeneous, coated, cylindrical specimen under external hydrostatic pressure (p_h), and if we denote the maximum, medium and minimum overall pressure acting on the sample S_1, S_2 and S_3, then

$$S_1 = S_2 = S_3 = p_h .$$

If the pore pressure (p_p) is non-zero, the concept of the effective stress may be expressed as

$$\tau_i = S_i - p_p , \qquad i = 1, 2, 3 ,$$

where τ_1, τ_2, τ_3 are the values of the maximum, medium and minimum effective stress acting on the specimen. If hydrostatic and uniaxial pressure are acting together,

$$\tau_2 = \tau_3 = p_h - p_p , \qquad \tau_1 = \Delta S + (p_h - p_p) ,$$

where ΔS is the differential uniaxial pressure. All the external pressures acting on the specimen skeleton are thus reduced to non-zero pore pressure, and the values of the effective stresses become the determining quantities in porous substances. Table 3 implies that the differences in the strength of the coated and uncoated specimens we observed may, in principle, be explained by introducing the effective stresses, although deviations were observed in some cases. A quantitative evaluation would, however, require a much larger number of measurements which, with the exception of the extreme pore pressures, would be made in a device in which the magnitude of p_p can be controlled. The values for some specimens may not agree too because the necessary condition of perfect pore interconnection in the whole volume of the specimen is not satisfied.

Figures 1 and 2 show the changes of the saturated remanent magnetic polarization (J_{rs}) due to hydrostatic pressure up to 200 MPa for specimens of two typical granodiorites treated in different ways. The comparison of the results of measuring all granodiorites yield the following two essential conclusions: There are substantial differences between the individual types in the stability of J_{rs} to hydrostatic pressure (in specimens 12 and 15 J_{rs} is relatively quite stable, whereas in specimens 1 and 19 the pressure of 200 MPa demagnetized more than 50% of the J_{rs}). These differences are mainly due to the properties of the ferrimagnetics the specimens contain, and are closely related to the spectra of the microcoercitive forces (see Table 1). Our measurements of the specimens of the same granodiorite, treated in different ways, indicate a more important fact. There are relatively small differences between the uncoated specimens, regardless of whether they were dried (a) or water-saturated (c) in their initial condition; the pressure variations of the remanent magnetic polarization in the coated specimens (b) always differ. These samples (with zero pore pressure) are much less stable to hydrostatic pressure. These differences are particularly conspicuous in granodiorites of low magnetic stability.

Considering that the samples saturated with the liquid medium do not display

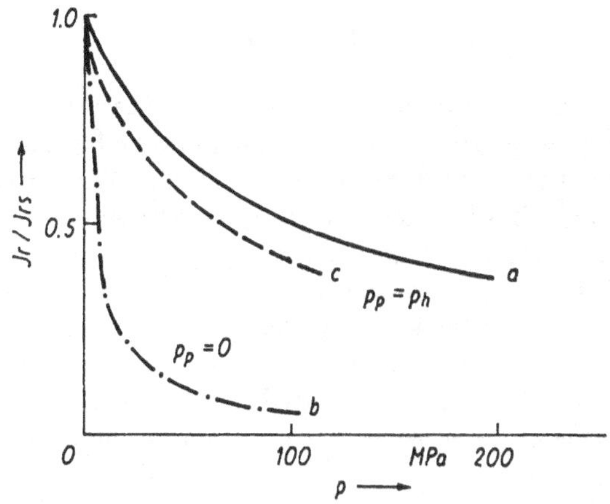

Fig. 1. Changes of saturated remanent magnetic polarization (J_{rs}) due to hydrostatic pressure in rocks with different pore pressure (p_p) — Specimen no. 1.

measureable changes in J_{rs} or initial susceptibility, the observed differences must be of purely mechanical nature and related to the differences in pore pressure. The remanent magnetic polarization of the specimens with zero pore pressure, and consequently with higher effective stress, is less stable than in the specimens in which the pore pressure is equal to the hydrostatic. This agrees qualitatively with the observed dependence of the strength of granodiorities on effective stress in samples with different pore pressures. Physically, the larger stability of J_{rs} in specimens, whose pores are filled by the liquid under pressure, is due to the change in the distribution of the internal stresses acting immediately on the ferrimagnetic minerals in the rock. Higher external pressures would be required to generated local stresses on the surfaces of these minerals, sufficiently large for demagnetization processes.

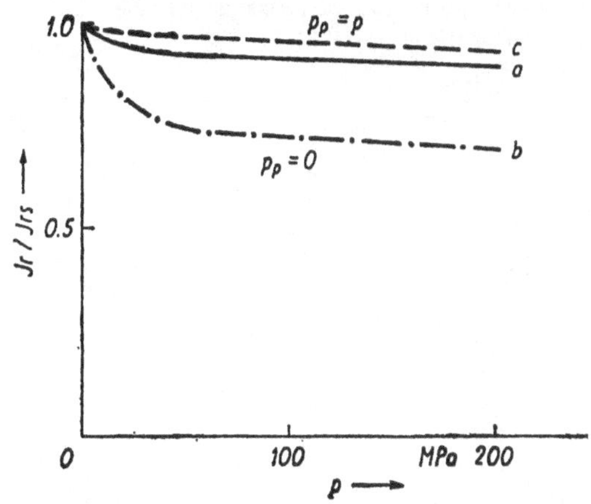

Fig. 2. The same as in Fig. 1 but for Specimen no. 19.

Conclusion

The magnetic and mechanical properties of rocks are determined, on the one hand, by the magnetic and mechanical properties of ferrimagnetic minerals and, on the other, by the process of transfer of external stress to these minerals. Whereas a number of theoretical and experimental studies has been devoted to explaining the former factor (summarized, e.g., in [6]), in which the fundamental physical mechanisms of interaction of the external stress with the ferrimagnetic mineral have been determined, much less attention has been devoted to the study of the transfer properties of the non-magnetic matrix. Our experiments have proved that the saturation of rocks with a liquid medium and different pore pressures have a substantial effect on their magnetic and mechanica properties, and that this effect must be taken into account in rocks with higher porosity.

References

[1] Ohnaka, M.; Kinoshita, H.: Effect of uniaxial compression on remanent magnetization. J. Geomag. Geoel. **20** (1968) 93–99.

[2] Kapička, A.; Savenko, B. J.: Comparative characterization of changes in basalt remanence under hydrostatic and uniaxial pressures. Geophysical Journal Ac. Sci. USSR, Kiev **1** (1979) 57–61.

[3] Nulman, A.; Shapiro, V.; Maksimovski, S.; Ivanov, N.; Kim, J.; Carmichael, R.: Magnetic susceptibility of magnetite under hydrostatic pressure and implications for tectonomagnetism. J. Geomag. Geoel. **30** (1978) 585–592.

[4] Handin, J.; Hager, R. V.; Friedman, M.; Feather, J. N.: Experimental deformation of sedimentary rocks under confining pressure: pore pressure tests. Am. Assoc. Petroleum Geol. Bull. **47** (1963) 717–755.

[5] Podroužkova, Z.; Mrazek, F.; Kapička, A.: Kamera kombinirovannoj nagruzki dlya deformatsionnykh charakteristik gornykh porod. Trans. Symp. High pressure technology, Liblice (1983) 124–129.

[6] Nagata, T.: Basic magnetic properties of rocks under the effects of mechanical stresses. Tectonophysics **9** (1970) 167–195.

CHAPTER V

Deformation Properties

Mechanical tests to determine rock strength characteristics

Miklós Gálos

Strength behaviour of rocks is described in terms of unidirectional compressive, tensile (indirect tensile) and triaxial test results. These tests deliver an ultimate value each of the rock strength characteristics, determined under a mechanical condition.

The mechanical condition comprises the system of assumptions for the rock as a material — namely that the rock is a homogeneous, isotropic, continuous material yielding a specimen that can be handled as an elementary unit, to result in an ideal stress state.

These tests have been standardized within the system of standards for rock materials, taking these assumptions into consideration:

MSZ 18285/1-79 Unidirectional Compressive Tests
MSZ 18285/2-79 Indirect Tensile Tests
MSZ 18285/3-79 Triaxial Tests.

Test results complying with these standards are of qualifying value, determinant in rock grading or in product quality testing.

Idealizing assumptions for structural materials cannot, however, be considered but as fair approximations. This is also clear from the fact that in tensile and compressive tests not all the specimen — hence the elementary unit — gets into ultimate condition at the instant of failure, but the material continuity fails at a point or minor part of the specimen, then the arising fault grows to a crack to give rise to the typical failure pattern.

The greatest part of an even failed — crushed, disrupted — specimen has not got into ultimate condition but keeps material continuity in some composite stress state.

In general, failure starts from a weak cross section, a structural fault inside the material.

These observations have led to the description of the failure phenomenon in mechanics with due consideration to the material texture, in agreement with the physical purport. The branch of mechanics concerned with failure is failure mechanics, relying for its development on research results by Griffith.

Within the failure process, Griffith distinguished arise and propagation of crack. He demonstrated by tests that in brittle materials — such as rocks — crack propagation is a decisive factor, since cracks (faults, pores) are a priori contained in the material, rather than to be produced by force effects.

Stresses at the tip of a crack of length $2c$, with a rouding radius r, can be written in closed form, analyzed according to the theory of elasticity. In in-plane stress state, for $r \ll 2c$ (Fig. 1):

$$\sigma_x = \frac{K}{\sqrt{2\pi R}} \cos\frac{\theta}{2}\left(1 - \sin\frac{\theta}{2}\sin\frac{3\theta}{2}\right),$$

$$\sigma_y = \frac{K}{\sqrt{2\pi R}} \cos\frac{\theta}{2}\left(1 + \sin\frac{\theta}{2}\sin\frac{3\theta}{2}\right),$$

$$\sigma_z = 0,$$

$$\tau_{xy} = \frac{K}{\sqrt{2\pi R}} \cos\frac{\theta}{2}\sin\frac{\theta}{2}\cos^3\frac{\theta}{2}.$$

Examining surroundings of the crack in in-plane deformation state, there is one difference:

$$\sigma_z = \nu(\sigma_x + \sigma_y),$$

where ν is the Poisson's ratio.

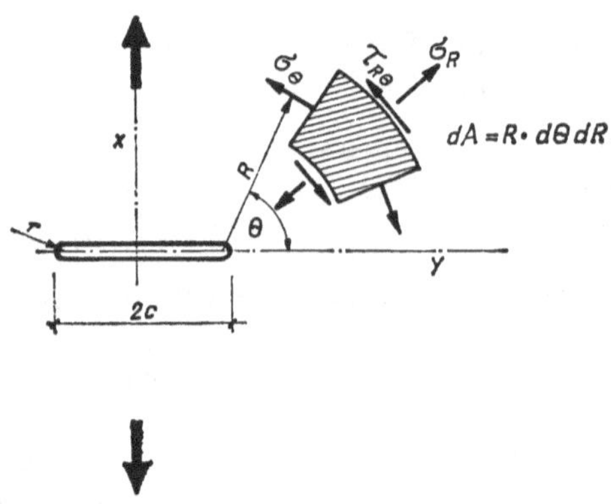

Fig. 1. Stresses at the tip of a crack of length $2c$.

These relationships unambiguously show that the value of any stress component arising in the surroundings of a crack tip — for given polar characteristics R and θ — is controlled by a single parameter, the stress intensity factor K.

This means also that the behaviour of the specimen (model) can be directly compared to that of a structure. In spite of different geometries, dimensions, loads, the cracks inside both systems behave similarly if K values are the same. From the aspect of characterizing the material, the critical case where the crack begins to propagate is of interest for us.

Thus, in our investigations, the critical value of stress intensity factor K entraining abrupt (unstable) failure is wanted. The critical stress intensity factor K_{Ic} is a material characteristic defining the load capacity of the material.

In any mechanical test to failure, it is a fundamental requirement to produce a mechanically clearly defined stress state in a notched specimen where the crack will start from the notch.

In case of metals, the crack starting from the notch is produced by fatigue testing. To do so would be frustrating for brittle materials such as rocks. It has been experienced that fatigue entrains failure before origin of a crack could be macroscopically observed. Therefore it is assumed for rocks that notching gives rise to the opening of a pore at the crack tip or to a microcrack at the mineral interface likely to initiate a crack starting from the notch root.

In analogy with steel tests, the three-point bending test has been applied to determine K_{Ic} in rocks, by making a notch in the middle of the bending test prism (Fig. 2). Thereby the test corresponds to the classic bending test. The relationship for the K value:

$$K = \sigma \sqrt{c} \cdot Y$$

takes the ratio of notch to specimen depths into consideration.

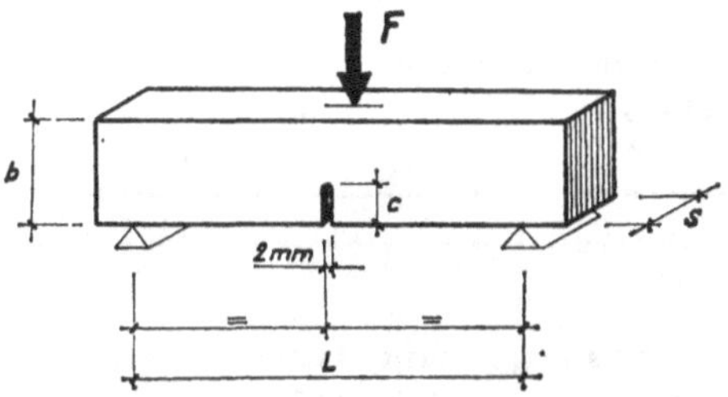

Fig. 2. The three-point bending test.

Hungarian standard MSZ 4927-76 specifying the three-point bending test of steel involves the determination of K_{Ic} as:

$$K = \frac{6FL}{sb^2} \sqrt{c} \cdot Y,$$

where: F — loading force (N), L — span between supports (mm), s — specimen width (mm), b — specimen depth (mm), c — notch with incipient crack, Y — correction factor.

Relying on a test series complemented with photoelasticity tests, the standard relationship for the correction factor:

$$Y = 1.99 - 2.47 \frac{c}{b} + 12.97 \left(\frac{c}{b}\right)^2 - 23.17 \left(\frac{c}{b}\right)^3 + 24.80 \left(\frac{c}{b}\right)^4.$$

K_{Ic} has been determined on two rock types by means of three-point bending tests, evaluating the effects of specimen geometry, and rock homogeneity.

Test results have been compiled in Tables 1 and 2.

Table 1. Examination of the stress intensity factor on a test prism of Andesite tuff (Szentendre)

Nominal specimen size (mm) L	s	b	Stress intensity factor K_{Ic} (N/mm 3/2)		
160	40	40		21.2	
				23.3	
			mean		22.3
160	20	40		22.8	
				21.3	
			mean		22.0
70	40	20		23.9	
				20.8	
			mean		22.4
70	20	20		18.9	
				20.7	
			mean		19.8
mean			$\bar{x} = 21.6$		
corrected standard deviation			$s = 1.6$		
number of tests			$n = 8$		
range			$\Delta = 5.0$		

Table 2. Examination of the stress intensity factor on a test prism of Compact limestone (Máriagyüd)

Nominal specimen size (mm) L	s	b	Stress intensity factor K_{Ic} (N/mm 3/2)		
160	40	40		31.9	
				36.2	
				28.8	
			mean		32.3
160	20	40		41.3	
				40.3	
			mean		40.8
70	40	20		20.5	
				37.8	
			mean		29.2
70	20	20		47.7	
			mean		47.7
mean			$\bar{x} = 35.6$		
corrected standard deviation			$s = 8.4$		
number of tests			$n = 8$		
range			$\Delta = 27.20$		

Andesite tuff (Szentendre) is a fine-grained rock of porphyric texture, containing weathered feldspars, that could be considered as homogeneous to the scale of testing, apparent also from standard deviations of test results. No low K_{Ic} values have been obtained but for specimens of very small cross sections.

Compact limestone (Máriagyüd) is a microcrystalline rock of carbonate texture interlaced with microcracks where specimens 20 mm deep did not yield realistic results, and even deeper specimens exhibited heterogeneity due to microcracks.

Our tests have established the fitness of three-point bending tests to determine K_{Ic} values as material constants also for rocks. In tests, homogeneity of the rock has to be reckoned with, and handled as an important material characteristic.

Further evaluation of the tests may follow after demonstrating existing relationships between test results and material characteristics, however, further tests have to be made.

Physical properties of rocks in the process of deformation under high pressures

E. I. Bayuk, N. I. Dyaur

The processes of rock deformation, taking place at various depths, are closely connected with such problems as oil and gas prospects at great depths, prediction of earthquakes etc. [1, 2, 3]. Laboratory investigations of changes of various rock properties in the process of deformation up to their destruction under high pressures permit to reveal physical characteristics, which can be put into the base to develop the precursors of rock destruction [3, 4]. Observation of physical properties becomes of great significance.

This paper shows the change of electrical resistance, velocity and amplitude of elastic waves in the process of sedimentary rock deformation, having various liquid pore pressures in complicated stress field.

It is known that in the Earth's crust, especially in its upper parts, the irregular three dimensional stress field occurs, which is defined by the values of mean normal stress and σ_i-stress, characterising the degree of stress irregularity. Stress state also depends on a liquid pressure in rock pores. Stress, being equal to the difference of the outer and pore pressures is called effective.

The Institute of geology and exploration of combustible fuels has developed the techniques of measuring the deformation, electrical resistance, velocity and amplitude of elastic longitudinal waves in the process of experiment under the influence of regular effective stress and took into account the liquid pore pressure and irregular stress, i.e. uniaxial compression under a high confining pressure. The known method of studying deformational characteristics of rocks [5] has been modified to establich the values of electrical resistivity or velocities and amplitudes of elastic waves at all values of stress with deformation alteration [6].

Simultaneous measurement of physical characteristics and installation of the sample with gauges are shown in Fig. 1. For deformation measurement the elastic tensometers have been used. Strain gauges for measuring the longitudinal and radial deformations are connected as the Wheatstone bridge. Its unbalances of indications coincide by time and they are recorded by two-coordinate recording instruments PDS-021 m. The same instruments record readings of the bridge diagrams for measuring the efforts in the process of tests under the irregular compression. In this case the record of stress-deformation diagrams takes place. Modified specially treated electrodes have been used for measuring the electrical resistance. These measurements have been done according to two-electrode circuit with a help of balanced four-pole bridge of alternating current. The velocity of elastic waves has been measured by ultrasonic impulse method within

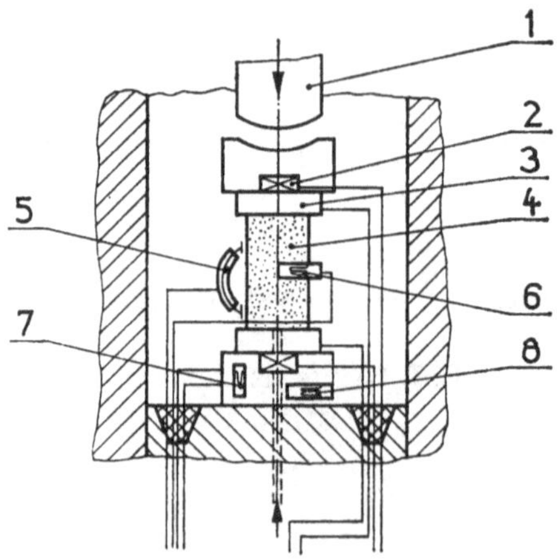

Fig. 1. Pattern of high pressure equipment. *1* — piston, *2* — piezoelectric pick-up, *5, 6* — strain gauges, *7, 8* — stress strain gauges.

frequency 0.8—1 MHZ. Two pairs of piezoelectric transducers have been installed on the faced surface of cylindric sample, having a diameter of 16 mm and a length of 25 mm.

The samples of polymict sandstones, tuffstones, limestones and dolomites have been selected for studies. Before the experiment the above samples were saturated with simulated formation water or kerosine. Pore pressure maintained constant during the experiment, being equal to hydrostatic pressure at the depth of sampling or greater. The increase of effective stress has been reached on the account of rising the confining

Table 1. Deformations (longitudinal — ε_1, radial — ε_2, relative electrical resistance — ϱ/ϱ_0 of rocks under pressures — σ_{eff}).

Rocks, number	Porosits (%)	σ_{eff} (Mpa)	ε_1 (%)	ε_2 (%)	ϱ/ϱ_0
Tuff-sandstone — 33	13.5	44	0.11	0.25	1.27
ferriferous-carbonatic cement		80	0.28	—	1.38
		110	0.33	0.37	1.42
Tuffstone — 65, carbonate cement	19.9	49	0.29	—	1.25
		80	0.42	—	1.37
Sandstone — 69, clayey-carbonatic cement	18.3	48	0.16	0.29	1.29
		69	0.21	0.35	1.45
Sandstone — 95, greywacke,	9.7	47	0.20	—	1.66
clayey-carbonatic cement		100	0.22	—	2.02
Sandstone — 96, carbonatic cement	16.6	48	0.20	—	1.32
		80	0.25	—	1.60
Sandstone-greywacke — 103	17.2	60	0.22	—	1.22
Tuff-sandstone — 100, carbonatic cement	16.7	68	0.32	0.60	1.71
Limestone — 70	9.2	49	0.15	0.25	1.12
		69	0.20	0.33	1.15
Domolite — 30	2.95	20	0.13	0.20	—
		70	0.22	0.32	—

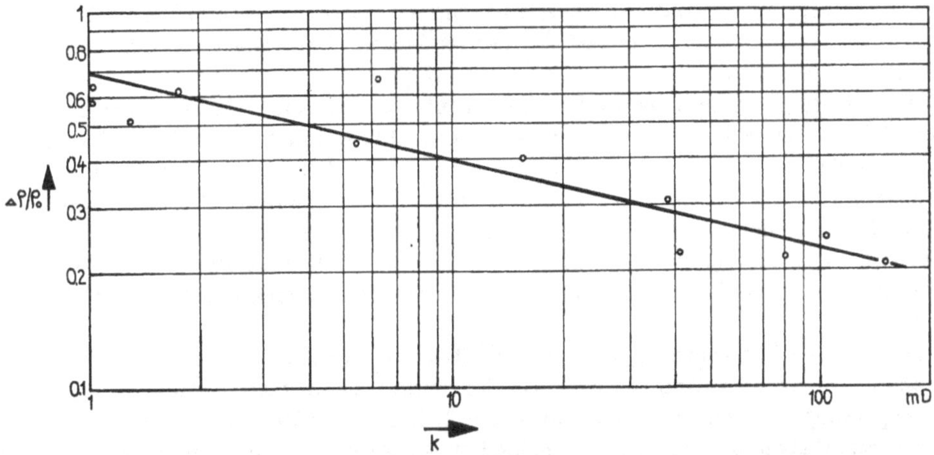

Fig. 2. Relative electrical resistance vs. permeability factor under 60 Mpa.

pressure. At each steps of stress the sample was kept up to stabilization of deformation and electric al resistance values. The experiments have been carried out at an ambient temperature.

Under the regular loading the effective stress increased up to a value corresponding to the depth of sampling or greater. Sometimes two samples of the same rock were tested up to different maximum, effective stress. The results of experiments are given in Table 1. The values of higher deformations are presented there: longitudinal — ε_1, radial — ε_2 and also relative electrical resistance — ϱ/ϱ_0.

Deformations and relative electrical resistance grow with increase of effective stress, The results on two samples of the same rock almost coincide, though some divergences. especially for the curves of relative electrical resistance have been observed. The electrical resistance is very sensitive to a structure of rock and is defined by winding of current conducting channels (microcracks etc.). Therefore even onconsiderable difference in micropore distribution in two samples of the same rock causes various changes of the relative electrical resistance under the influence of effective stress. A low changes of ϱ in limestones in comparison to the observed one in sandstones should be mentioned. On the whole the relative alteration of electrical resistance is closely related to the rock permeability. In more permeable rocks the growth of electrical resistance with pressure increase is less (Fig. 2). Such rocks preserve good filtration properties at great depths under hydrostatic pressure conditions.

Zones of anomalous changes of deformation, relative electrical resistance are typical for the majority of rocks (Fig. 3a). It is connected with structural alterations of rocks under thermodynamic conditions, corresponding to the occurence depth [7]. However the effective stress under which the anomalies take place, usually surpasses a stress at the depth of occurence by 10, 20 and more Mpa. In some experiments anomalous changes have not been traced, though the effective stress increased up to 110 Mpa (sandstones 33, 95), as there is a relationship with rock formation environment and lithogenesis processes. If there is a predominance of clastic volcanogenic rocks in these sandstones, then at a slight deformation change the rapid increase of electrical resistance is observed as a result of current conducting channels alteration under a pressure.

The velocity of longitudinal waves in liquid saturated rocks under the influence of hydrostatic compression increases slowlier [8, 9] and it grows even slowlier under the rise of pore pressure.

When testing the samples under the irregular compression the obtained graphs of the relative electrical resistance — stress considerably complete the deformation graphs. Such graphs at different stresses are given in Fig. 3b. When deformation changes lineary with stress the electrical resistance increases or maintains almost constant. The above fact can be traced in the elastic zone. While transmitting into the nonelastic zones the electrical resistance begins to drop. The strength is reliably defined according to both graphs. Both the strength and the flow limit increase with effective stress growth.

The values of strength σ_{ic} and flow limit σ_{is} obtained at the mentioned effective stresses in tested rocks and the values of deformations and relative electrical resistance, corresponding to them, are given in Table 2. Limestone-70 has the highest strength properties. Its residual deformation, being low at a stress corresponding to the occurence depth (49 Mpa), highly increases (more than 3 times) when the outer pressure rises by 20 Mpa. Sandstones-96 and 33 have high values of strength properties if the effective stresses are 70—110 Mpa and a residual deformation increases at that.

Polymict sandstone with clayey carbonatic cement-69 and tuffstone-100 had low strength properties. Like limestones these rocks appeared to be restrictly plastic at an occurence depth and their plasticity grows with the increase of depth.

The electrical resistance — stress can be defined according to diagram. Such rocks are predisposed to shattering and discompaction under the irregular compression. It is

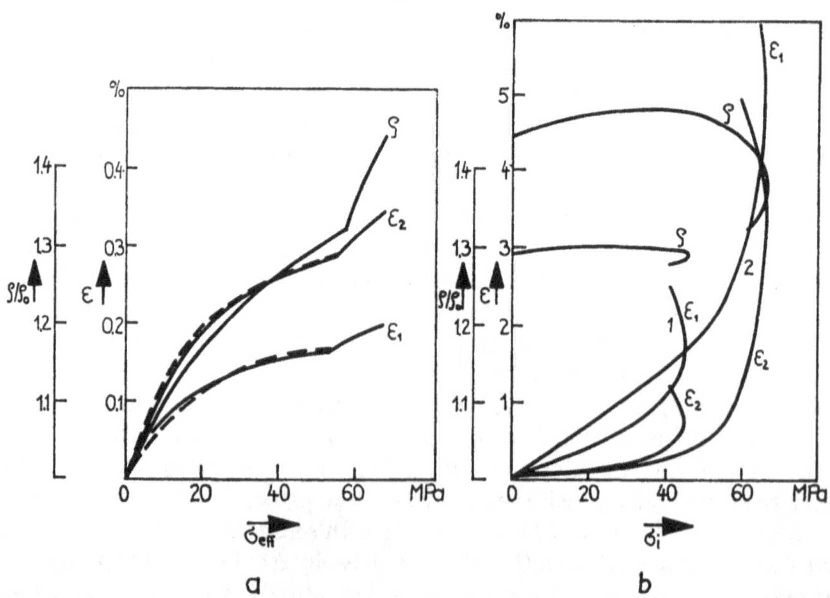

Fig. 3. ε_1 — longitudinal, ε_2 — radial deformation and relative electrical resistance, ϱ/ϱ_0 vs. effective stress, σ_{eff} — a) and axial stress, σ_i b) for two sandstone samples, tested under effective stress of 49 Mpa (1) and 69 Mpa (2).

Table 2. Deformations and strength of rocks at different stress states

Rocks, number	σ_{eff} (Mpa)	σ_{is} (Mpa)	ε_1 (%)	ε_2 (%)	ϱ/ϱ_0	σ_{ic} (Mpa)	ε_1 (%)	ε_2 (%)	ϱ/ϱ_0
Tuff-sandstone — 33	80	65.3	0.18	0.09	1.440	121	1.46	0.83	1.04
	110	128.0	—	0.17	1.325	184	—	5	1
Tuffstone — 65	80	34.5	0,25	—	1.372	138	2.85	—	1.10
Sandstone — 69	48	28	0.54	0.09	1.295	43.9	1.75	0.76	1.29
	68	42	1.50	0.48	1.480	69.2	5.08	3.84	1.38
Sandstone — 95	100	51	0.50	—	1.485	150	5	—	1.2
Sandstone — 96	48	68	0.78	—	1.500	169	2.75	—	1.47
	80	95	0.60	—	1.595	180	5.23	—	1.4
Sandstone — 103	60	97	0.75	—	1.260	160	1.80	—	1.23
Tuff-sandstone — 100	68	32	0.35	0.10	1.66	64	—	—	1.36
Limestone — 70	49	54.4	0.17	0.08	1.120	175	1.90	0.70	0.82
	69	71.4	0.25	0.08	1.142	200	6.55	5.03	0.78

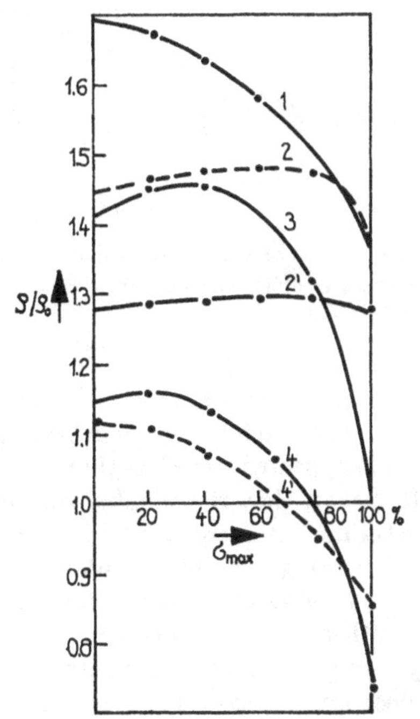

Fig. 4. Alteration of relative electrical resistance — ϱ/ϱ_0 under axial stress (in percentages of maximum). *1* — tuffstone-100 under outer effective stress of 68 Mpa; *2* — polymict sandstone-69 under 69 Mpa; *3* — the same under 48 Mpa; *4* — tuffstone-33 under 110 Mpa; *5* — limestone-70 under 69 Mpa; *6* — the same under 49 Mpa.

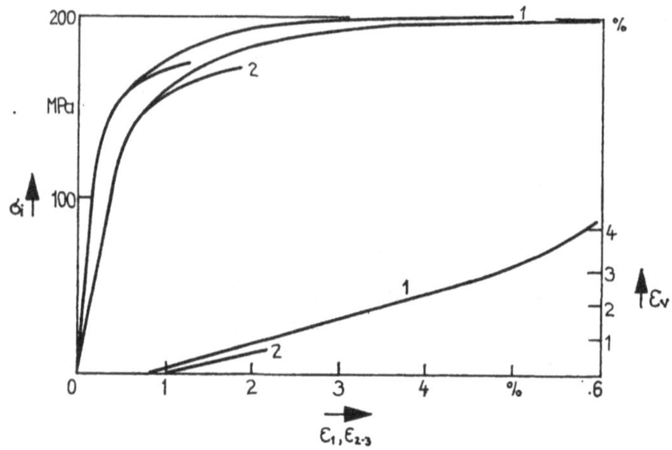

Fig. 5a)

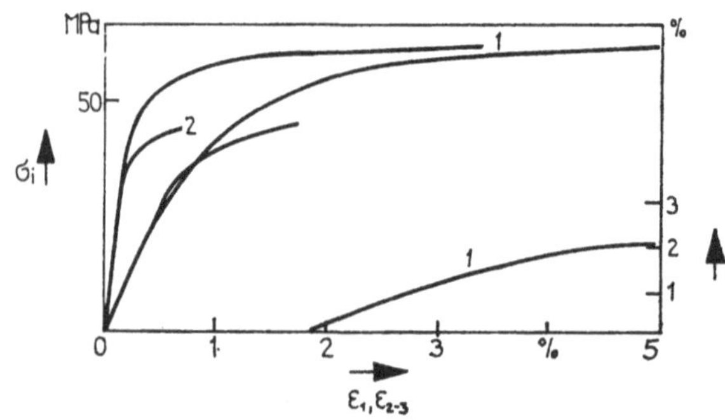

Fig. 5b)

Fig. 5. Deformational curves and curves of volume deformation alteration for limestone-70 and sandstone-69b. *1* — under outer effective pressure of 69 Mpa; *2* — under 49 Mpa.

well shown in the graph of relative resistance vs. percentage of maximum compression stress σ_{max} relationship. Such a graph for a number of rocks is given in Fig. 4. In some rocks the electric resistance decreases so intensevely in non-elastic zones that at a stress corresponding to the strength it appeared to be less than the initial value, i.e. the values at atmospheric pressure. It means, that the number of current conducting channels increased. Hence, microfracturing has increased and rock filtrating properties have improved. For limestone-70 shattering takes place under the effective stress corresponding to the occurence depth, and for sandstones-33, 96 the above process occurs under higher effective stresses. In other rocks, for example in sandstones-69, 100, 103 the electrical resistance slightly changes, i.e. maintains rather high even at the moment of destruction. Permeability does not increase under a complicated stress state or it is considerably low.

The deformation of limestone-70 and sandstone-69 are compared in Fig. 5. They have different strength properties but close values of longitudinal deformations. Under the outer pressure of 49 Mpa the volume of limestone increased by 1% and there was no increase of sandstone volume because of radial deformation difference. Under 69 MPa the volume growth in limestones is twice that in sandstones. It agrees with the data on the electrical resistance.

The example of velocity and amplitude alteration of the longitudinal waves in sandstones under a complicated stress state is given in Fig. 6. The outer pressure was 120 MPa and the graphs of relative alterations of longitudinal deformation, velocity and amplitude of longitudinal waves vs. compression stress relationship were given in the percentages of its maximum value. The velocity and amplitude in the zone of elastic deformation initially sharply increase and when the formation of microfractures begins predominating over their closing the degree of growth somehow decreases. In the zone of non-elastic deformations the velocity value gradually decreases. Just before the fracturing the velocity of longitudinal waves sharply drops.

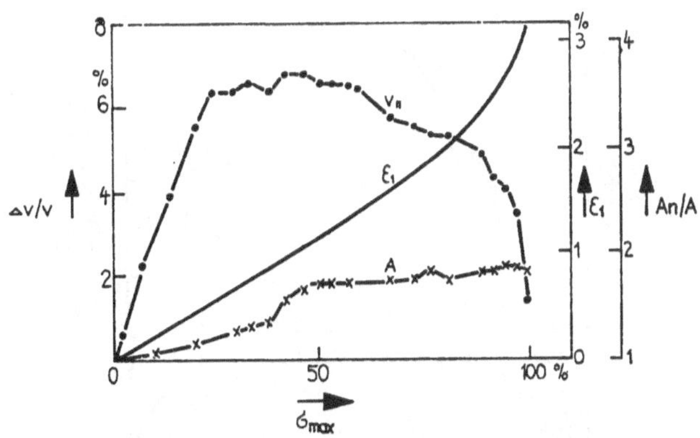

Fig. 6. Relative alteration of longitudinal deformation — ε_1, velocity — v and amplitude — A of longitudinal waves vs. stress (in percentages of maximum) — σ_{max} for sandstone under 120 MPa.

Comparing Fig. 6 with electrical resistance alteration in sandstones under the influence of maximum failure stress one can reveal, that in the zone of elastic deformation the electrical resistance rises slowlier than the velocities of longitudinal waves. Non-elastic deformation causes more intensive drop of the electrical resistance than the velocity. The above process takes place as the formation of microfractures and just before the rock destruction promotes the decrease of environment effective elasticity and the velocity of longitudinal waves stops rising and then it drops under the formation of principal fracture. Primary development of fractures, filled by water, considerably improve the electric conductivity at the initial stage of deformation and the electric resistance begins to drop.

Let us consider the influence of saturation on the velocity of longitudinal waves under a complicated stress state. Previously only the effect of liquid pore pressure on the velocity under an hydrostatic compression has been studied [9].

The samples of dolomite-30 saturated with kerosine have been tested under the axial compression and outer pressure of 30 and 120 MPa, the pore pressure being 10 and 50 MPa, accordingly. In dry samples of dolomite the velocity under 30 MPa was equal to 4.25 km/sec and under 120 MPa — 5.22 km/sec. The velocity of longitudinal waves in samples saturated with kerosine under the same pressure was almost not distinguished, ranging from 5.24 to 5.27 km/sec for different samples.

The velocity and volume deformation alterations for dolomite-30 caused by the stress (in percentages of the failure stress) are shown in Fig. 7. The data on non-saturated dry samples, being deformed under 30 MPa and the data on saturated samples with pore pressure of 10 MPa, also being deformed under 30 MPa, have been correlated (Fig. 7a). The similar data under 120 MPa and pore pressure of 50 MPa are given in Fig. 7b. It is clear that the velocity alteration in dry samples strongly depends on the outer pres-

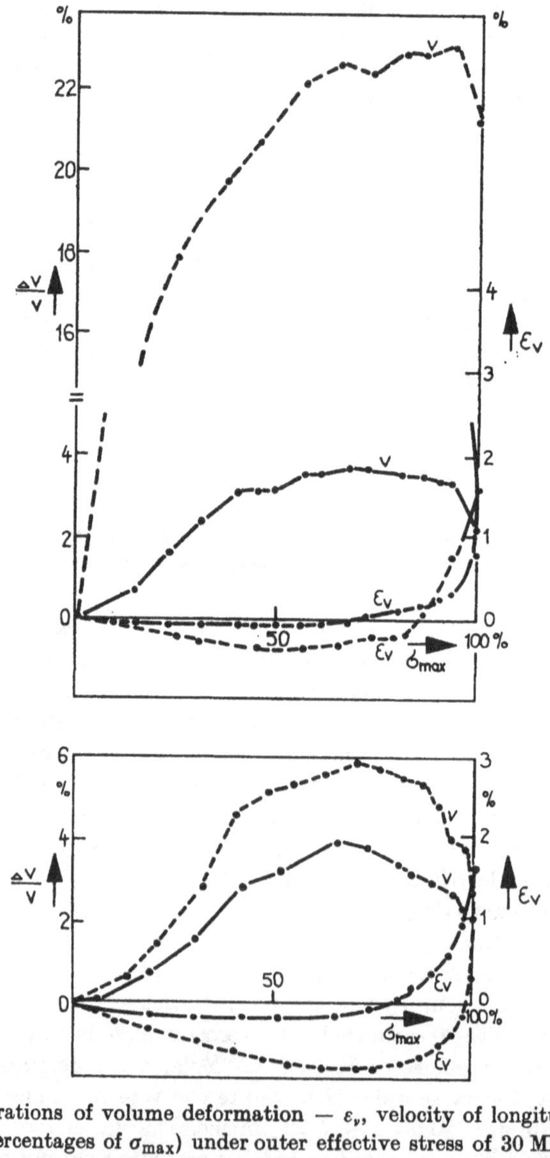

Fig. 7. Relative alterations of volume deformation — ε_v, velocity of longitudinal waves — v vs. dolomite stress (in percentages of σ_{max}) under outer effective stress of 30 MPa and pore pressure of 10 MPa — a) under outer pressure of 120 MPa and pore pressure of 50 MPa — b). *1* — dry sample; *2* — saturated with kerosine.

sure. The velocity of longitudinal waves in the range of elastic deformation under 30 MPa increases almost by 23% and under 120 MPa rises approximately by 6%. The velocity changes slightly (50—95%), reaching the maximum value, and then just before the destruction it sharply drops.

On the whole the nature of velocity dependence on stress maintains for the saturated rocks with different pore pressure. It means that the zones of velocity increase, slight alteration and drop can be distinguished. Even only rock liquid saturation leads to a considerable velocity increase, therefore the additional velocity growth in the process of sample deformation is 4%. There is an essential difference between the curves for dry and saturated rocks, but the influence of pore pressure on the velocity of longitudinal waves in dolomite is poorly expressed. One can recognize only more extended zones of the velocity alteration, that is the zone of stable values under lower pore (10 MPa) and confining (30 MPa) pressures.

Analysing the results of the volume deformation under a stress it is possible to establish that the compaction of a saturated dolomite in the elastic zone is 4—5 times as less as a dry one. However, the presence of pore pressure accelerated the beginning of dilatension, which develops under 5—10% σ_{max} stress and in dry samples under 35—55%. Dilatensive discompaction ($\varepsilon_v > 0$) also started earlier. Such a type of volume deformation alteration under a stress in sedimentary samples with high pore pressure permits to suppose that in natural conditions of the Earth's crust pore pressure may promote the preservation or even the improvement of reservoir properties of rocks, especially in tectonically active zones.

Conclusions

1. The technique of complex studying the physical properties of sedimentary rocks under a complicated stress state and different pore pressure of saturating liquid has been developed.

2. By the alteration of complex physical characteristics in the process of rock destruction three zones have been distinguished: in the zone of elastic deformation the velocity and amplitude of longitudinal waves rise, the electrical resistance slightly grows and the deformations linearly change. With the development of dilatension and with the formation of fractures the values of velocity and amplitude of longitudinal waves are stabilizing, the electrical resistance starts decreasing. Before the destruction the velocity of longitudinal waves and electrical resistance sharply decreases and, as a rule, the volume deformation increases (discompaction).

3. The increase of pore pressure strengthens the effect of discompaction. And the electrical resistance changes more than the velocity of longitudinal waves.

Reference

[1] Physica ochaga zemletrjaseny. Ed.: M. A. Sadovsky. Nauka (1975) 244 p.
[2] Neftegazonosnost bolshich glubin. Nauka (1980) 119 p.
[3] Prognoz zemletrjaseny. Ed.: Donish. Dushanbe-Moskva (1983) 4.

[4] Volarovich, M. P.; Tomashevskaja, I. S.; Budnikov, V. A.: Mechanica gornich porod pri visokich davlenijach. Nauka (1979) 151 p.
(5] Pavlova, N. N.: Deformationny i collectorskie svoistva gornich porod. Nedra (1975) 240 p.
[6] Bayuk, E. I.; Fomin, A. A.; D'aur, N. I.: Izuchenie deformirovanija ocadochnich porod po izmeneniju physicheskich characteristic. In: Physika ochaga i predvestniki zemletrjaseny. Nauka (1981) 2.
[7] Dobrynin, V. M.: Deformaty i izmenenija physicheskich svoistv collectorov nefti i gaza. Nedra (1970) 239 p.
[8] Bayuk, E. I.: Vlijanie davlenija na skorost uprugich voln v nasishennich gidkostjami obraztach ocadochnich porod. In: Physichesky svoistva collectorov nefti pri visokich davlenijax i temperaturach. Nauka (1979) 31—37.
[9] Avchjan, G. M.; Matveenko, A. A.; Stefankevich, Z. B.: Petrofisika osadochnich porod v glubinnich uslovijach. Nedra (1979) 244 p.

CHAPTER VI

Complex Investigations

Physical properties of heated samples of fine-grained sandstone from coal measures

A. Fojtek, J. Švec, P. Martinec, M. Krešl

1. Introduction

The investigated sandstone belongs to those rocks which surround most frequently coal seams in the Czechoslovakian part of the Upper Silesian Coal Basin. These rocks have a porosity lower than 3%, a high uniaxial compression strength (110 to 140 MPa) and a tensile strength of 8 to 9 MPa. The objective of our experiment was to determine changes in selected physical parameters of samples of sandstones subjected to heating up to 900 °C.

2. Petrography of samples

In our experimental work, samples of sandstone belonging to the Poruba member of Ostrava formation, Namurian A were used. It is a grey, well graded, uniformly grain-sized sandstone of a massive structure with unpronounced subhorizontal bedding.

The content of clastic quartz and stable rock fragments is 49% by volume, the content of unstable fragments and mineral grains is 25% by volume and the matrix is 24% by volume.

The clastic phase of the sediment is represented by subangular quartz grains, fragments of quartzite, sericitic quartzites and aggregares of quartz-feldspar grains from acid magmatites. Unstable fragments are represented by the phyllites, mica schists and sediments. The biotote is chloritized and baueritized. It prevails over the clastic muscovite. The contact of the grains of the clastic phase is of the convex-concave type with regeneration of clastic quartz grains. The mean diameter of the clastic quartz grains is 0.15 to 0.23 mm, maximum grain size of quartz is 0.53 mm.

The matrix is quartz-sericite-clay with a small amount of carbonate cement. The clay minerals are represented by illite and chlorite. The coal matter is detritic, disseminated in the rock. In the quantitative system of clastic rocks proposed by J. Konta (1972), this sandstone corresponds to fine-grained lutitic greywacke.

It has been deduced from the chemical analysis that it is a rock with a high ratio of SiO_2 and Al_2O_3 (5.38) and K_2O and Na_2O (1.67). Together with mineralogical composition, these ratios characterize the rock as relatively stable from the point of view of the geochemistry. The diagenesis of the sediment corresponds to the stage of late catagenesis as described by Longvinenko and Kossovska [1]. There is association of phylosilicates of illite and chlorite with total volume of pores equal to 0.03588 $cm^3\,g^{-1}$ and a low absorption capacity.

3. Preparation of samples

The samples were prepared by cutting the cores of a diameter of 48 mm into individual specimens. The bedding of the rock may be deduced from the deposition of mica scales and of anisometric clastic grains which are normal to the axis of the drilling core.

The velocity of longitudinal acoustic waves, thermal conductivity, electrical resistivity and dielectric constant were determined. After these measurements, the specimens were heated for 4 hours in oxidation atmosphere in the interval of 200 to 900 °C with increments of 100 °C. After the heating, the investigated physical properties were determined once more on the specimens as well as the weight loss.

The specimens heated to 500 °C and 800 °C could not be used for the measurement of physical properties because owing to a hidden primary fracturing, the heating induced a secondary fracturing in them.

4. Methods of measurement

Physical properties of basic (not heated) specimens were measured on specimens dried during 48 hours at a temperature of 105 °C. The heated samples were put into dessiccator. The measurement of physical properties was made after the cooling of the specimens to room temperature.

The loss of weight in % by weight (Δm) of the initial specimen was determined by measuring the weight before and after the heating. In addition to it, the loss of weight was determined from the curve of weight loss of powdered sample recorded during a parallel recording of the curve of differential thermical analysis in oxidation atmosphere at the heating rate of 10 K min^{-1} in the interval of 20—1000 °C.

The velocity of propagation of longitudinal acoustic waves v_p was measured using the EMA apparatus which makes it possible to measure the duration of the passage of the ultrasonic wave through the investigated sample. For these measurements, samples of a diameter of 48 mm and a length of 50 mm were used.

The thermal conductivity was determined using the method of test bar. Teflon was taken as standard. The specimens had the form of cylinders of a diameter of 48 mm and a length of 50 mm.

The electrical resistivity ϱ was measured using the bridge method with the bridge type BM 559 (manufactured by Metra Blansko). The specimens had the form of cylinders of a diameter of 48 mm and a length of 20 mm. The measurements were taken in the dessiccator.

The dielectric constant ε_r was measured by a comparative method using the bridge BM 559 (by Metra Blansko) [2]. Teflon was used as standard. The specimens had the form of circular plates of a diameter of 48 mm and a thickness of 5 mm. The measurements were taken in a dessiccator.

5. Results of measurement

The Tables 1 to 6 show the results of measurements (i.e. mean values from 4 to 5 measurements). Graphical representation of these results is shown in Figs. 7 to 11.

Table 1. Loss of weight Δm of heated samples of sandstone (determined by measuring the weight before and after the heating of the samples)

Heating temperature	[°C]	100	200	300	400	500	700	900
Δm	[%]	0.35	0.3	0.4	0.85	2.2	2.85	3.0

Table 2. Loss of weight Δm of heated samples of sandstone (determined from the TG curve)

Temperature °C	Δm %	Temperature °C	Δm %
115	0.5	535	1.35
160	0.75	575	2.05
225	0.8	620	2.6
325	0.95	635	2.75
380	1.05	680	3.0
425	1.2	720	3.25
470	1.35	740	3.3
		845	3.45
		1000	3.6

Table 3. Velocity of propagation of longitudinal acoustic waves v_p through heated samples of sandstone ($m \cdot s^{-1}$)

Temperature of heating °C	Velocity of longit. acoustic waves v_p	
	before heating	after heating
200	3270 ± 30	3270 ± 20
300	3280 ± 20	3280 ± 30
400	3280 ± 10	3280 ± 20
600	3320 ± 30	2480 ± 10
700	3250 ± 30	1770 ± 10
900	3300 ± 20	1520 ± 10

Table 4. Thermal conductivity of heated samples of sandstone ($W \cdot m^{-1} \cdot K^{-1}$)

Temperature C°	Thermal conductivity	
	before heating	after heating
200	2.38 ± 0.14	2.50 ± 0.12
300	2.39 ± 0.17	2.55 ± 0.16
400	2.44 ± 0.14	2.34 ± 0.16
600	2.42 ± 0.16	1.89 ± 0.14
700	2.47 ± 0.16	1.69 ± 0.11
900	2.35 ± 0.19	1.27 ± 0.09

Table 5. Electrical resistivity of heated samples of sandstone (Ωm)

Temperature of heating °C	Electrical resistivity	
	before heating	after heating
200	5200 ± 500	5830 ± 600
300	5500 ± 800	10210 ± 1000
400	6000 ± 1000	18150 ± 1500
600	6000 ± 700	64000 ± 7000
700	5400 ± 500	255000 ± 40000
900	5900 ± 400	717000 ± 85000

Table 6. Dielectric constant of heated samples of sandstone

Temperature of heating °C	Dielectric constant	
	before heating	after heating
200	7.8 ± 0.3	7.7 ± 0.5
300	7.5 ± 0.4	7.2 ± 0.6
400	7.8 ± 0.4	7.7 ± 0.6
600	7.6 ± 0.7	6.5 ± 0.5
700	7.9 ± 0.7	6.5 ± 0.6
900	7.7 ± 0.8	5.3 ± 0.6

6. Discussion of results

6.1. Changes in composition and texture of heated samples of rocks

The following conclusions could be drawn from the analysis of changes in mineralogical composition and texture of the rocks as a result of their heating:

1. Up to the temperature of 140 °C, water freely bonded in the pores and adsorbed on the surface of the sample was liberated. The composition and the texture of the rocks remained unchanged. That condition of the rock applied up to the temperature of 200 °C (Fig. 1 — SCAN 1).

2. When heated to 300 °C, the samples exhibited an initial destruction of the dispersed coal matter. Even though the content of coal matter in the investigated sample was low (less than 2%), its destruction induced changes in the texture of the rock.

3. In the interval of 300 to 400 °C, a beginning of destruction of the chloritized biotite was observed in the oxidation atmosphere, as well as de-hydroxylation of illite and further destruction of the coal matter (Fig. 2 — SCAN 2).

4. In the intercal of 400 to 700 °C, the above reactions continued, i.e. the de-hydroxylation of the illite and the destruction of the chlorite. Owing to a different thermal expansion of the matrix and the clastic phase (quartz), microcracks were formed at the boundaries of the grains already at the temperature of 400 °C. Such a cracking of the rock was even more pronounced after having reached the temperature of 572 °C as a

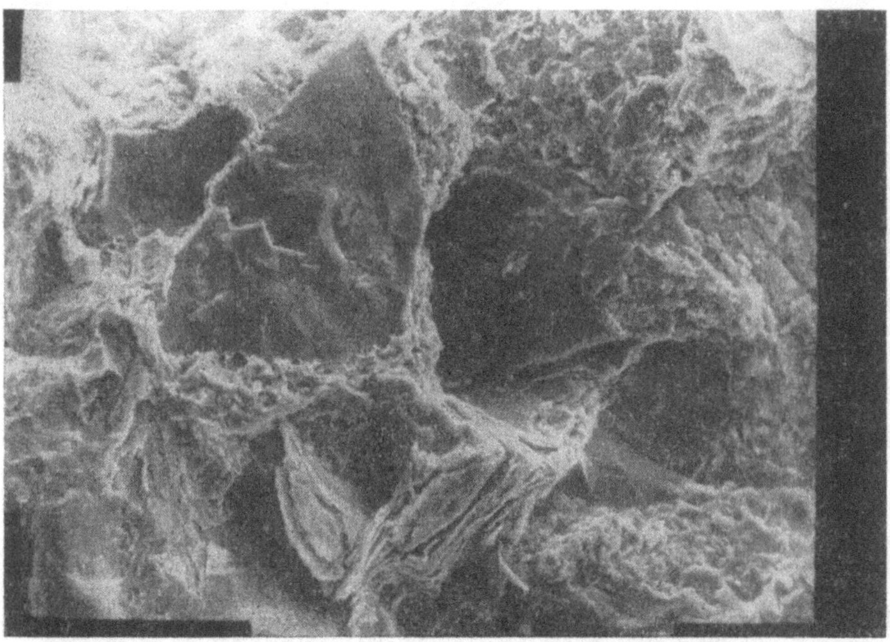

Fig. 1. Scanning electron microscopy. Original sample of sandstone.

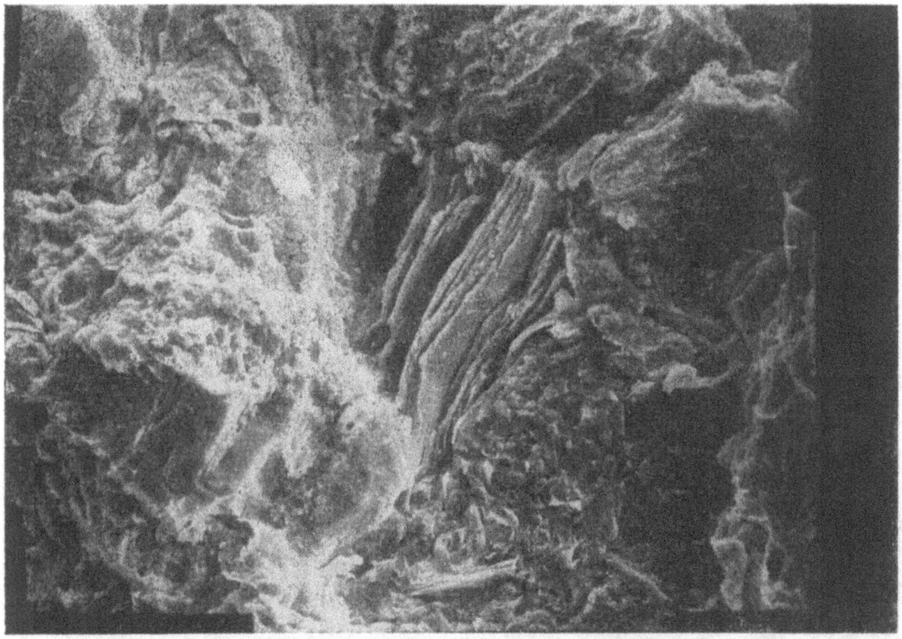

Fig. 2. Scanning electron microscopy. Sample heated to 400 °C.

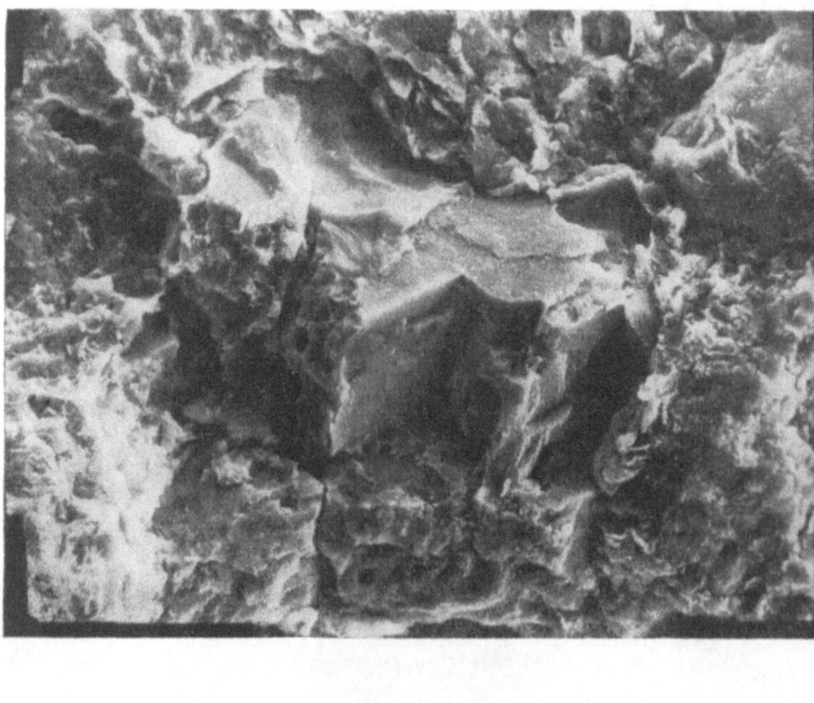

Fig. 3. Scanning electron microscopy. Sample heated to 600 °C.

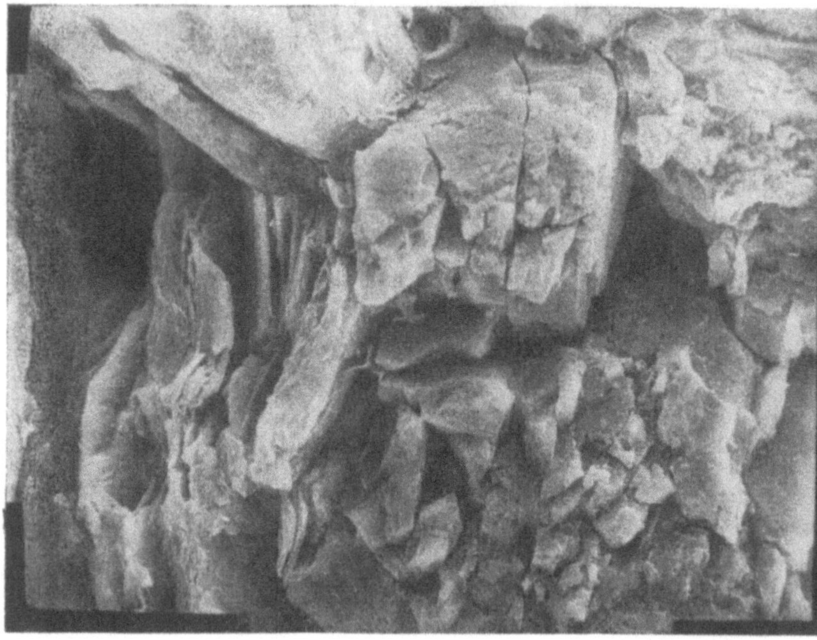

Fig. 4. Scanning electron microscopy Sample heated to 700 °C.

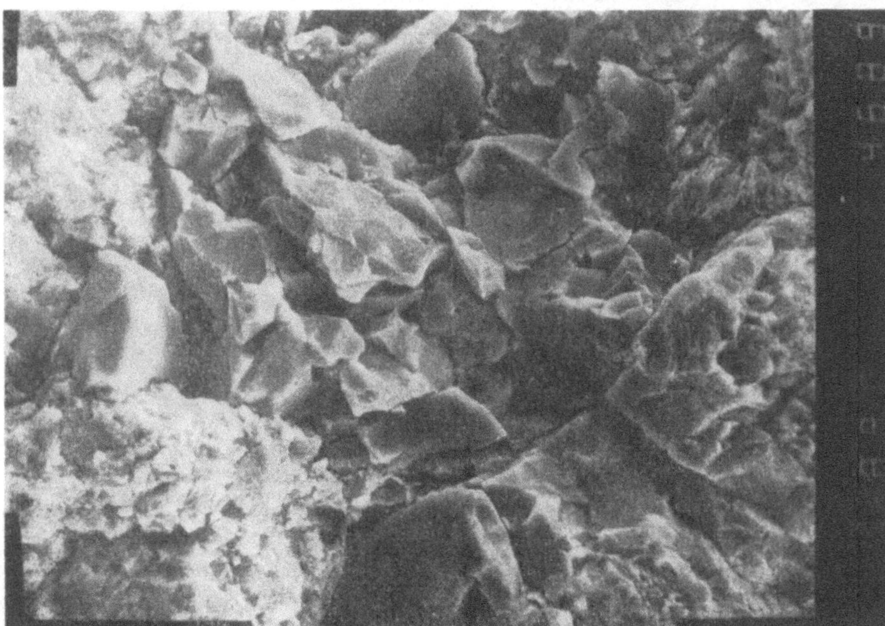

Fig. 5. Scanning electron microscopy. Sample heated to 900°C.

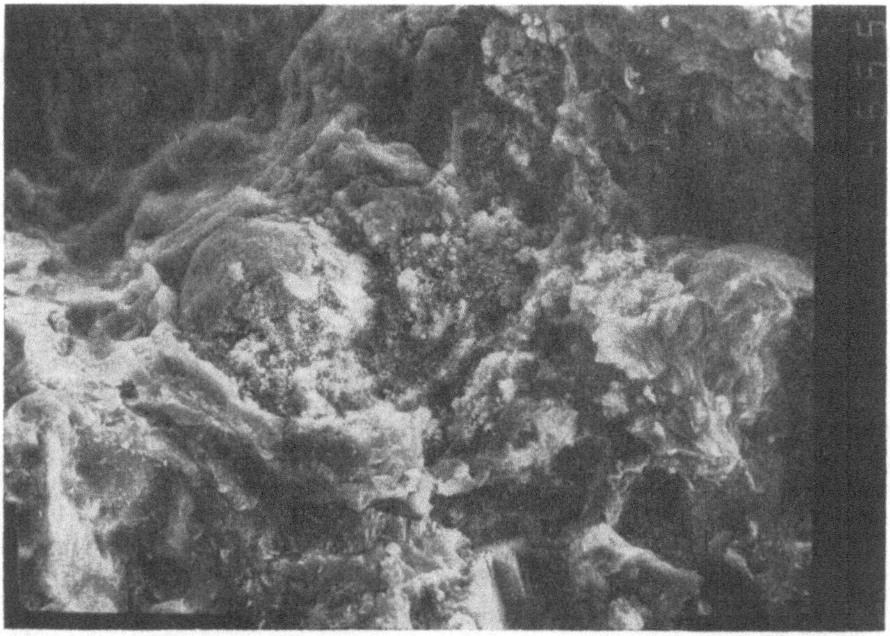

Fig. 6. Scanning electron microscopy. Sample heated to 900°C.

result of changes in the volume of the quartz induced by the transformation of $\alpha \rightarrow \beta$ quartz. The cracking was also observed in clastic grains of quartz and in fragments of rocks rich in quartz (Fig. 3 — SCAN 3). In the region of 600 °C, the chloritized biotite was decomposed completely and its products were oxidized. The scales of illite exhibited higher interference colours and greater fineness of grains.

5. In the region of 700 °C, the oxidation could be observed even macroscopically in integral samples, i.e. the products of the decomposition of chlorite and chloritized biotite had a brown colour. The clay matter exhibited a beginning of isotropization and particles of illite caused a reduction of interference colours. The boundaries between grains were more pronounced and the cracking proceeded transversally along the clastic grains (Fig. 4 — SCAN 4).

6. The sample heated to the temperature of 900 °C had a white colour, with red and brown points visible macroscopically. The thermally unstable minerals were recrystallized, small quantities of carbonates were dissociated. The phase of spinel appeared. Intense fracturing of the rock was observed both in the matrix and in the clastic grains. Frequent conchoidal fractures occurred in quartz grains as well as polygonal decomposition of original integral grains (Fig. 5, 6 — SCANS 5, 6).

6.2. Loss of weight

Curves corresponding to the loss of weight of the samples are shown in Tables 1 and 2 and in Fig. 7. The differences between the slope of the curve $\Delta m = f(t)$ obtained from thermal analysis and from integral heated samples are due to the mode of heating (in thermal analysis 10 K min^{-1}, while in integral sample the heating took place 4 hours at the given temperature) and to the character of the sample (powder-integral sample). The following regions may be delimited at these curves:

a) the region of dehydration of the sample (in thermal gravimetric curve up to 150 °C, in heated samples up to 100 °C),
b) the region of relatively stable weight up to the temperature of 380 °C,
c) the region of dehydroxylation of illite, destruction of chlorite and chloritized biotite with oxidation of reaction products,
d) the region delimited in c) passes continuously into the region of dissociation of ankerite-dolomite up to the temperature of 850 °C.

Over the temperature of 700 °C, the changes in weight are insignificant.

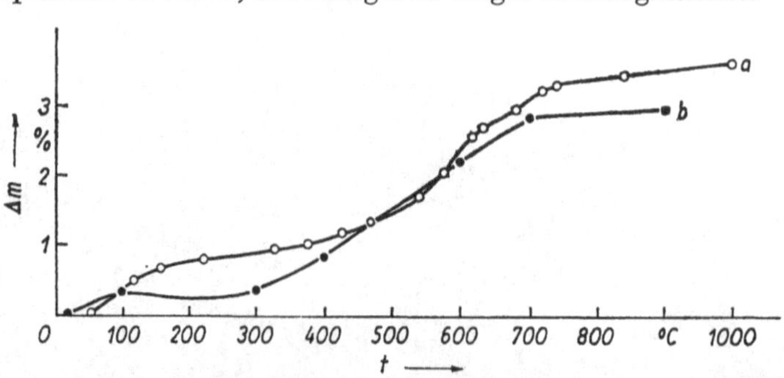

Fig. 7. Loss of weight (Δm) of heated samples
—o— thermal gravimetric analysis; —·—·—·— heated integral samples.

6.3. Velocity of acoustic longitudinal waves

It may be deduced from the diagram shown in Fig. 8 that:

- in the investigated temperature interval, total reduction of velocity of longitudinal acoustic waves is 50%,
- up to the temperature of 400 °C, the velocity of acoustic waves in the heated samples changes in very small amounts,
- the decrease of the velocity was observed in the interval of 400 to 700 °C as a result of microcracks produced in the samples due to different thermal expansion of clastic grains and matrix,
- the formation of spinel phase (700 to 900 °C) provokes a retardation of the decrease of the velocity of acoustic waves owing to a new cementation of the texture of the rock in which the temporarily stable mineral associations are excluded.

The changes in the velocity of longitudinal acoustic waves passing through the heated samples of sandstone reflects both structural and textural characteristics of the rock.

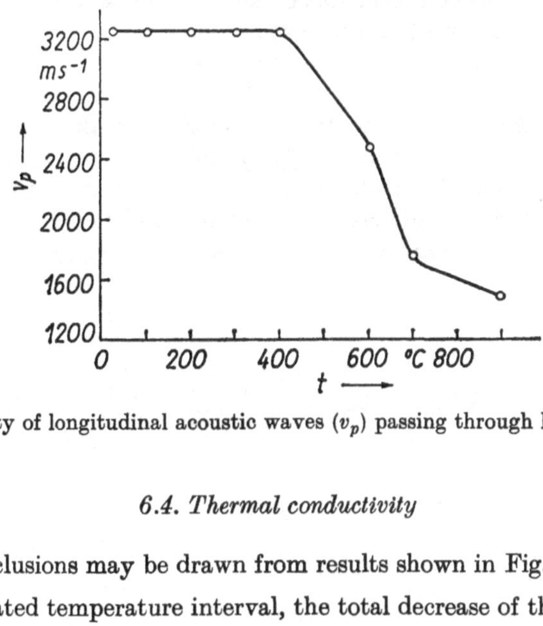

Fig. 8. Velocity of longitudinal acoustic waves (v_p) passing through heated samples.

6.4. Thermal conductivity

The following conclusions may be drawn from results shown in Fig. 9:

- in the investigated temperature interval, the total decrease of thermal conductivity is 55%,
- in heating interval of 200 to 400 °C, the thermal conductivity changed insignificantly, which is in an agreement with the character of the texture and composition of the rock,
- above the temperature of 300 °C, the thermal conductivity of the heated samples decreases linearly owing to non-reversible changes in the samples (i.e. disturbance of the texture of the rocks by the effect of changes in the volume of quartz grains and in fragments rich in quartz which are the main components of the rock, and also by the effect of non-reversible mineral transformations intensified by the oxidation of some of the reaction products).

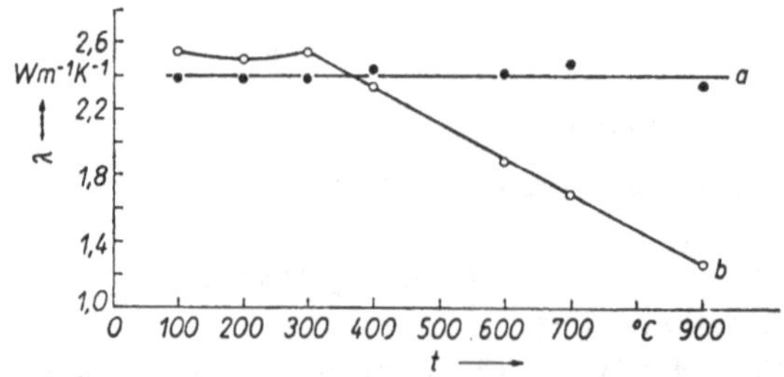

Fig. 9. Thermal conductivity (λ) of heated samples.
a) mean value of basic, unheated samples; b) heated samples.

6.5. *Resistivity*

The dependence of resistivity on the temperature of heating (Fig. 10) may be characterized in the following way:

- starting from the temperature of heating of 300 °C, the resistivity exhibits an increase, which is due to non-reversible mineralogical changes (changes in the texture of micas and clay minerals) and to textural changes (microcracks and disturbance of the sample as a result of differences in thermal expansion of clastic grains and the matrix),
- a pronounced increase of resistivity was observed at the temperature of heating over 600 °C which may be interpreted as a manifestation of the gradual formation of the spinel phase (substitution for the matrix) and as a manifestation of disturbance of the texture of the rock and clastic quartz grains as well as stable fragments. The fractures formed in them are a result of transformation of $\alpha \rightarrow \beta$ quartz.

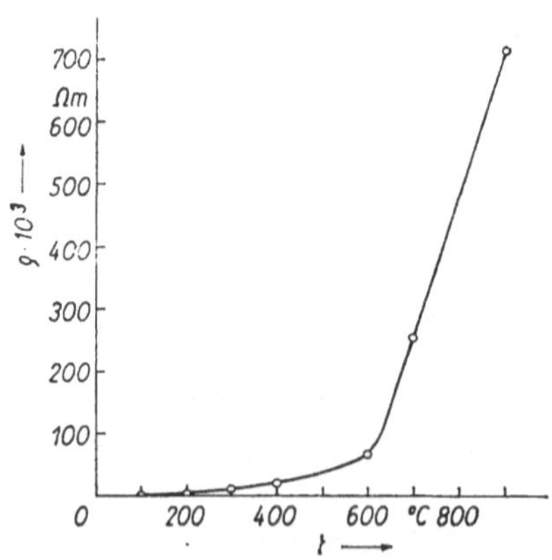

Fig. 10. Resistivity (ϱ) of heated samples.

6.6. *Dielectric constant*

The dependence of dielectric constant on the temperature of heating (Fig. 11) may be interpreted in the following way:

- in the investigated temperature interval, the dielectric constant decreased some 33%,

— in the interval of 200 to 400 °C, the dielectric constant changed in a very small amount,
— over the temperature of 400 °C, a gradual decrease of the dielectric constant was observed as a result of nonreversible textural changes in the heated samples (first of all due to the cracking of the samples by the effect of different thermal expansion of the clastic phase and the matrix of the rock).

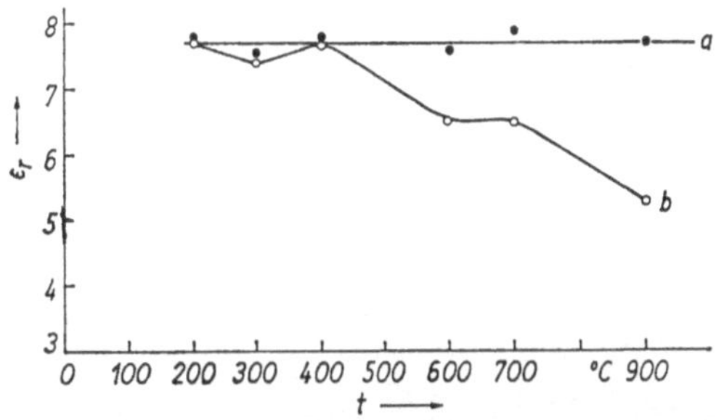

Fig. 11. Dielectric constant (ε_r) of heated samples.
a) mean value of basic unheated samples; b) heated samples.

7. Conclusions

Owing to mineralogical and textural changes in the finegrained sandstone under investigation, heated up to 900 °C, the following conclusions could be drawn from the experiment:

— From 20 to 400 °C, the mineralogical composition is conserved, even though there is a decomposition of the coal matter and a release of the freely bonded water. The disturbance of the texture of the rock owing to the differences in thermal expansion of clastic grains and the matrix was not great enough to induce changes in physical parameters of the rock.
— From 400 to 700 °C, important changes in the minerals and an intense disturbance of the texture of the rock owing to changes in the volume during the transformation of $\alpha \to \beta$ quartz were observed. The physical pattern of the rock is modified significantly: there is an increase of the loss of weight, a decrease of the thermal conductivity and of the velocity of longitudinal acoustic waves, an increment of the electrical resistivity and a reduction of the value of the dielectric constant.
— From 700 to 900 °C, a new texture of the rock with anhydrous mineral associations, which are stable even after the decrease of the temperature to room temperature, was formed gradually. The changes in the weight were insignificant, the decrease of the velocity of acoustic waves was delayed, the thermal conductivity as well as the dielectric constant decreased and the electrical resistivity increased abruptly.

References

[1] Longvinenko, N. V.: Postdiageneticheskie izmenenia osadochnyh porod. Nauka, Leningrad, 1968.
[2] Vašinek, V.; Švec, J.: Dielektriká konstanta karbonských hornin. In: Sborník 3. mezinárodní vědecké konference, sekce 3, VŠB, Ostrava, 1985.

Investigation of connections among grains of magmatic rocks

FRIGYES EGERER

Introduction

Mineralogists and specialists in physics of solids are mainly interested in the character of the existing connections among the mineral grains constructing the rocks, the question of surface layers as well as the settlement of charges on the surface and at the grain boundaries. This subject was investigated by Zoltai, Stout, Newnham, Egerer and many other specialists.

General problem

A certain part of problems dealing with the connections among grain boundaries can be summed up as follows. Let us take a well crystallized rock, a granite of high solidity as an example. It is a well-known fact that if we make a thin section (for example a section 20 μm thick) and go on making it thinner and thinner, the polished sample will break in one or more places. These fractions occur very rarely at the grain boundaries, but they especially frequently go through the grains.

Petrographers investigating physical properties of rocks observed and pointed out, that the resistance to the electric flow and the flow of heat and other energies is stronger at the grain boundaries than it is inside the grains. In the case of certain minerals the decomposition was observed as a process starting not on the lateral face of the prism, but on the open transversal surface and taking place during the fracture of the prism, where it became considerable. Such observations have been taken into account by Zoltai, Wojnarovits and other authors.

Comparing the distribution of ions and consequently that of the charges we can conclude that they are different at the grain boundaries and on the surface of the minerals from the distribution inside the grains and minerals.

Experiments and results

The electron microprobe is an extremely suitable medium for following with attention the elements and the charges belonging to them. We are going to introduce a series of analyses made in a granite.

In Fig. 1, you can see a photo of two grains: one of an orthoclase and one of a quartz.

In Fig. 2, point 1 you can see the photo of an X-ray spectrum, measured on the orthoclase; in point 2 we are showing the same X-ray spectrum measured on the quartz grain. It can be seen clearly that the grains measured were really grains of an orthoclase and that of a quartz.

Fig. 3 shows the distribution of silicon ions in space. It can be observed very well in the photograph, that the quantity of silicon decreases at the grain boundaries. The distribution of potassium ions can be observed in Fig. 4. The decrease of the potassium

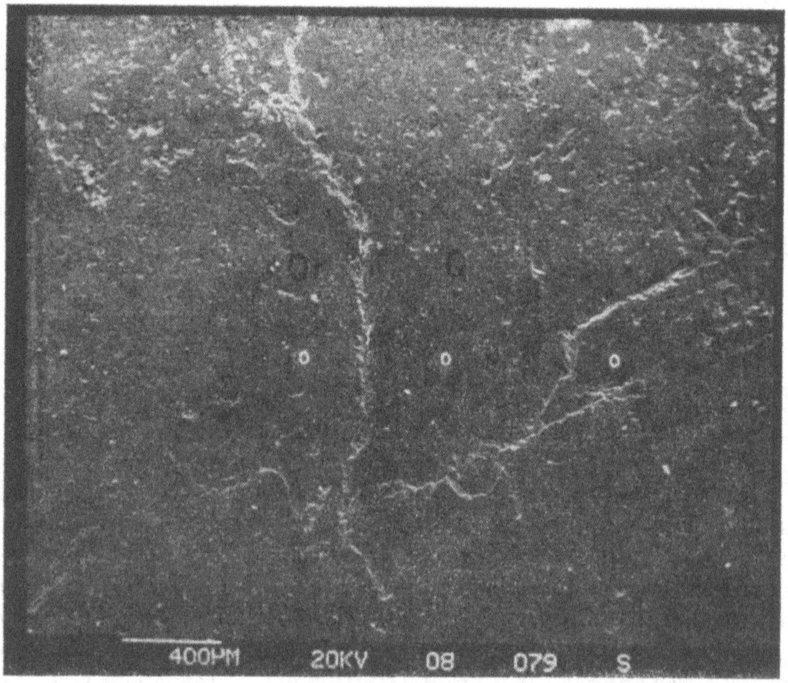

Fig, 1. Two grains: one of an orthoclase and one of a quartz.

quantity can also be detected in this figure. The same phenomenon can be seen in the "line-scan" photographs of the potassium in Fig. 5. Similar decrease can be found in the case of aluminium ions in the "line-scan" photo, shown in Fig. 6.

The distribution of other observed elements was investigated on the basis of local X-ray spectrum by a similar method. Elements occurring on the surface of the sample were the following: Si, Al, K, Fe, Zn, Cr, Na, Ca, Ti, Hg and S.

The "line-scan" photographs of the elements occurring in the sample are shown together, in the surroundings of the grain boundaries in Fig. 7.

The curves are photos taken with different intensity. In order to create a visual comparison, the representation was also made with different intensity. The values of the relative intensity can be seen beside the curves in Fig. 7. If we compare the curves in Fig. 7, we can make the following statements. The quantity of the main elements, the orthoclase and quartz are composed of — that is the quantity of potassium, silicon and aluminium — shows a great decrease at the grain boundaries, as well as in the zone among the grains. The quantity of sodium being present as a contamination also decreases. At the same time the quantity of certain elements and cations decreases considerably in

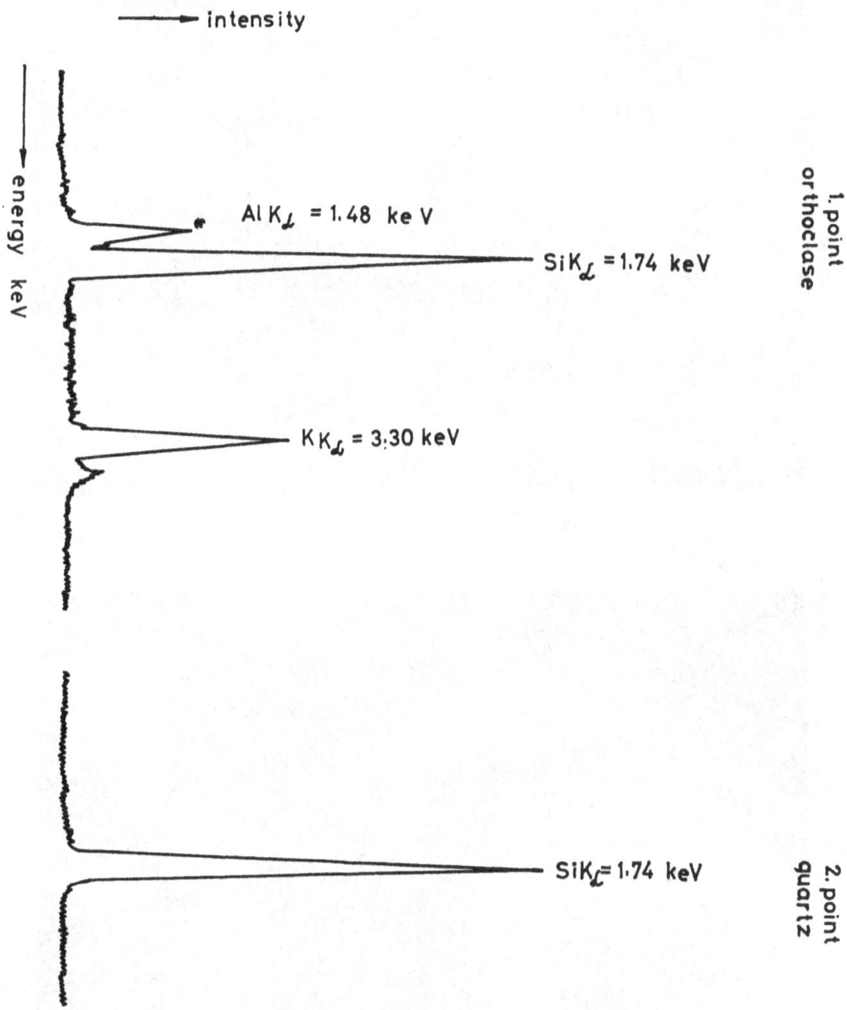

Fig. 2. X-ray spectrum of the orthoclase.

the zone as well as at the nearby grain boundaries. Such elements are, first of all, chromium and iron, but the same phenomenon can be observed to a certain extent at the boundaries in the case of zinc, quicksilver and sulphur.

Comparing the above mentioned examples with investigations of eruptive rock samples it is clear that the quantity of the main elements forming the mineral grains (K, Si, Al, Na, etc.) decreases very much in the zones between the grains and at the

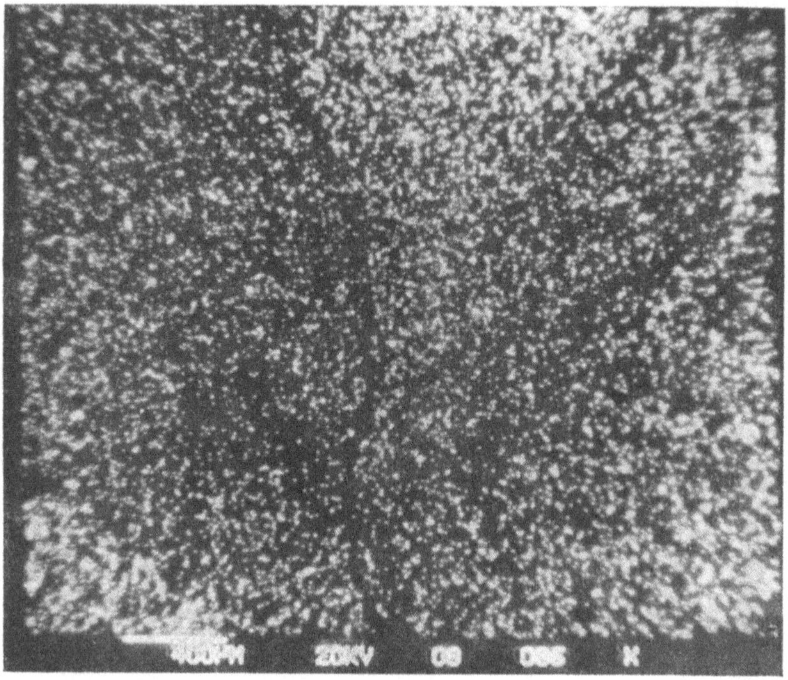

Fig. 3. Distribution of silicon ions.

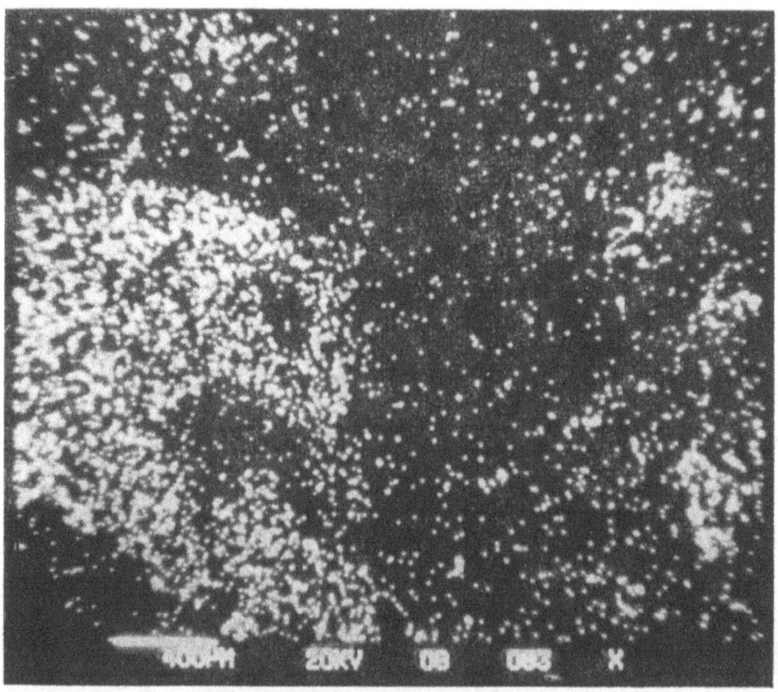

Fig. 4. Distribution of potassium ions.

Fig. 5. "Line-scan" photograph of potassium.

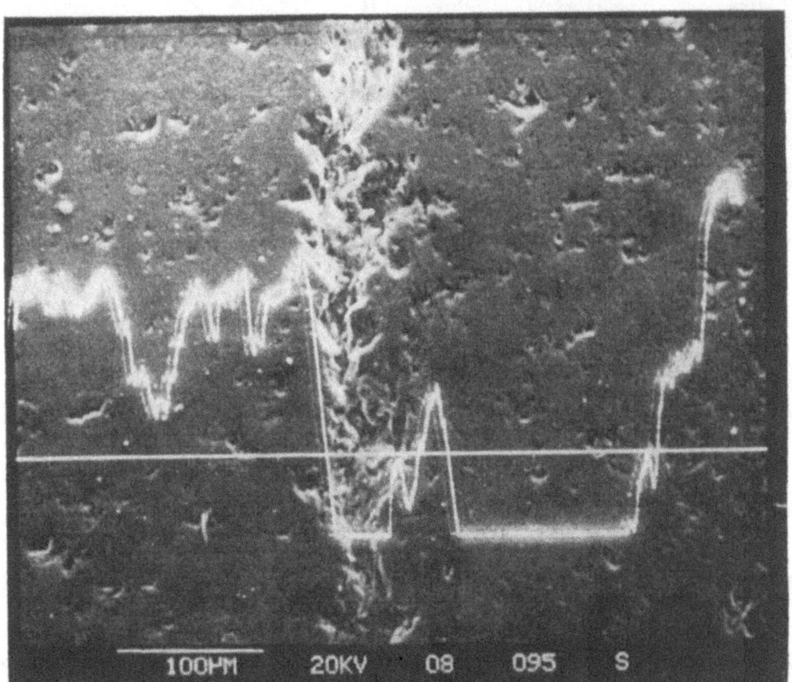

Fig. 6. "Line-scan" photograph of aluminium.

boundaries, and it is possible to substitute them by elements having the same serial number, such as iron, chromium, etc. It means that the composition of the chemical compounds in the zones among the grains and at the boundaries is greatly different from the chemical compounds existing inside the grains.

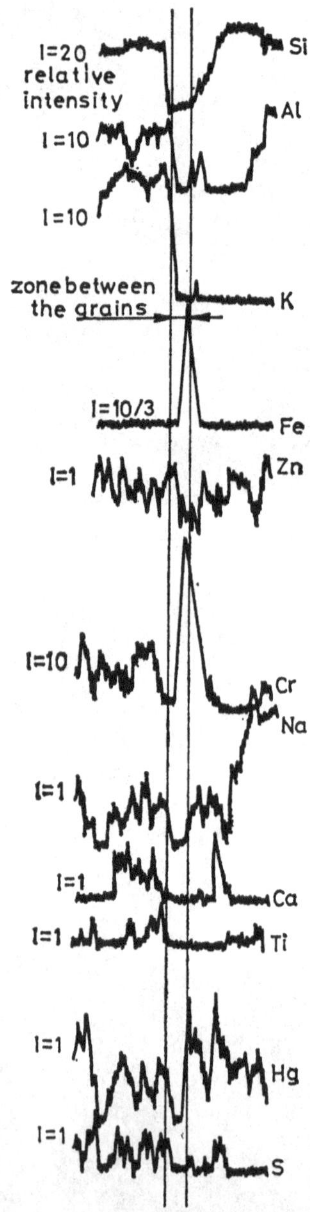

Fig. 7. "Line-scan" photograph of elements in the surroundings of the grain boundaries.

Results and classic theory of crystallization

As a result of works by Bowen, Goldschmidt and others the theory of the process of crystallization can be followed with attention in many reference-books, such as the works by Rankama and Kalervo.

The general course of crystallization of calc-alkalic basaltic magma is as follows:

Early magmatic stage:	Silicates (dunite, anorthosite) Sulfides (pyrrhotite-pentlandite paragenesis) Oxides (ilmenite, chromite, magnetite)
Main stage:	Gabbros Diorites Granites
Late magmatic stage:	Pegmatites Pneumatolyc deposits Hydrothermal deposits Volcanic emanations.

According to the introduced theory oxides and sulphides separate out in the early magmatic stage and the most frequent cations of this process are iron and chromium. It is proved by our results that the great majority of these two ions can be found at the grain boundaries and in the zone between the grains. But it is extremely difficult to make this very fact consistent with the conception that the oxides and sulphides of iron and chromium are formed in the early magmatic phase, and they are followed by the separation of silicates only in the main phase.

It is difficult to suppose — even if we disregard the differences caused by the consequent differences in density — that sulphides and oxides of iron and chromium granulate and form a "net" and later the holes in the net are filled by the silicates.

Conclusions

The classical theory of crystallization was not proved by our experiments, but moreover our results are in contradiction to it. That's why we are entitled to suppose, that the bonds at the grain boundaries must greatly differ from the bonds inside the grains.

It is probable that the result of these anomalous bond relations is that the charges are more strongly bound to the structure than they are tied inside the grain. This leads to the fact that the resistance to the flow of all kinds of energies (electric, heat, etc.) increases, and the mechanical fractures caused by the stronger bonds mentioned in the introduction are more frequent in the grains than at the boundaries.

Summary

The flow of charges and energy as well as the formation of mechanical fractures are investigated by means of an electron microprobe. The author of this paper points out that the classical theory of crystallization contains contradictions having not been discussed till now.

References

CHELIDZE, T. L.: The electric spectroscopy of heterogeneous systems. Kiev (in Russian) (1977).

ZOLTAI, T.; STOUT, J.: Mineralogy, Concept and Principles. Burgess Publ. Minneapolis (1984).

NEWHAM, R. E.: Domains in minerals. Am. Miner., **59** (1974) 906–919.

EGERER, F.: Kőzetfizika Tankönyvkiadó Budapest (1977).

PARHOMENKO, H. I.: Electrical Properties of Rocks, Plenum, New York (1967).

LEBEDEV, T. S.; SHEPEL, S. I.: Thermobaric studies of electric parameters of granitoids (in Russian). Geofiz. Zh. **6** (1984) 4, 14–23.

WOJNAROVITS, I.: Behavior of glass fibers in strong acid and alkali media. J. Amer. Ceram. Soc. **66** (1983) 896–898.

RANKAMA, K.; SAHAMA, T. G.: Geochemistry. Chicago Press. Chicago and London (1964) 161 p.

CHAPTER VII

Phase transitions in Earth materials

Phase changes in the upper mantle transition zone

W. Ullmann

Summary

The dimensionless parameter Δ, originally defined to be the negative gradient of specific entropy with depth z times the Grüneisen parameter and the quotient of temperature T and gravity acceleration, is tantamount to the relative deviation of the effective compressibility from that which would occur at z under isentropic conditions. In the latter form, Δ can be computed through the PREM model. The greater-than-one ratio C_P/C_V of the two principal specific heats C_P and C_V proves to be in excess of $\Delta + 1$ at every z.

Comparatively large deviations of Δ from zero arise in the upper mantle transition zone between $z = 220$ km and $z = 670$ km. The positive values of $\Delta(z)$ for $400 \leqq z$ [km] $\leqq 600$ entailing excessive values for C_P/C_V support the assumption that phase transitions occur in that depth range. Thus, according to PREM, the discontinuities at $z = 400$ km and $z = 600$ km are due to a phase change of first and second order, respectively. Other phase transitions occurring between those boundaries must be of second order.

The slope $\delta P/\delta T$ giving the rate at which the pressure P must change with temperature for the two alternative phases at depth z to remain in equilibrium proves to be negative at $z = 400$ km and, in contrast to this, positive at 600 km. Hence it follows that the elevations and depressions of the two principal phase changes within the transition zone are negatively correlated. Moreover, further phase changes occurring in the range $400 < z$ [km] < 600 feature positive values for $\delta P/\delta T$.

Introduction

The present analysis of thermal features within the upper mantle transition zone, TZ, is closely connected with the Earth model called PREM [Dziewonski and Anderson, 1981]. Hence our special attention shall be focussed on the region intermediate between 400 and 600 km depth. It should also be mentioned that TZ reaches from 220 km down to 670 km depth at which TZ is separated from the lower mantle by a discontinuity surface.

It is rather unlikely that any of the transient regions within the Earth is in fact perfectly flat. Therefore its representation as a spherical surface at well-defined depth is neither more nor less than an approximation to the reality. The only two discontinuity surfaces at 400 and 600 km depth to be taken into consideration are supposed to be suitable approaches.

The seismic velocities v_p and v_s as well as the density ϱ, each of them considered as a function of depth z in TZ, are continuous throughout, apart from the depth $z = 400$ km at which they undergo a fairly pronounced jump. The other discontinuity surface within TZ originates only from the gradients of the field functions v_p, v_s and ϱ carrying out a jump at $z = 600$ km, and therefore it is comparatively weak.

High-pressure experimental work suggests minerals undergoing phase transitions to higher-density structures at the presumable conditions of TZ. There is every appearance that much of the change in physical properties observed across the discontinuity surfaces at 400 and 600 km, and possibly other interfaces intermediate between them, is due to the different mineral assemblages existing on each side of these surfaces. Such a discontinuity surface could be separating two ranges of different chemical composition that are out of equilibrium. Alternatively, the discontinuity could be caused only by a phase change in an otherwise homogeneous medium.

However, neither of these alternatives can be strongly favoured on the basis of currently available data. For details reference may be made to the comprehensive paper by Jeanloz and Thompson [1983] in which the new information available on the phase equilibria and physical properties of minerals that occur within the mantle is reviewed [see also Lees, Bukowinski and Jeanloz, 1983; Ringwood, 1982]. In the following we are guided by the assumption that the intermediate layer of TZ is a region of phase changes. Thus each discontinuity surface belonging to the closed depth interval $400 \leqq z\,[\text{km}] \leqq 600$ will be considered an equilibrium structure that corresponds to reactions among the mineral phases at high pressures and temperatures.

The parameter Δ reconsidered

In a recent paper (Ullmann 1984a) the dimensionless parameter

$$\Delta := -\gamma \frac{T}{g} \frac{\mathrm{d}S}{\mathrm{d}z} \tag{1}$$

has thoroughly been discussed. The quantities T, S, γ and g are the temperature, specific entropy, thermal Grüneisen parameter and acceleration of gravity, respectively. This defining equation is to correlate the gradient $\mathrm{d}S/\mathrm{d}z$ with Δ as a scalar field which is, unlike $\mathrm{d}S/\mathrm{d}z$, completely computable on the basis of PREM.

To realize the last we note that (1) can be transformed into

$$\Delta = \frac{\varkappa_S}{\varkappa} - 1, \tag{2}$$

where $\varkappa_S$ and $\varkappa$ respectively are the isentropic and effective incompressibility at depth z [Brown and Shankland, 1981]. The latter is defined by $\varkappa := \varrho(\mathrm{d}P/\mathrm{d}\varrho)$. Using the hydrostatic fundamental equation $\mathrm{d}P/\mathrm{d}z = g\varrho$, we have $\varkappa = g\varrho^2/(\mathrm{d}\varrho/\mathrm{d}z)$. The computability of $\varkappa_S$ for every z is due to the relation $\varkappa_S = \varrho\Phi$, where: $\Phi := v_p^2 - (4/3)\, v_s^2$ is the seismic parameter.

It should also be mentioned that PREM yields v_p, v_s and ϱ in the form of linear polynomials in z for each of the three subranges of TZ. Hence, bringing to bear the Poisson equation for computing g, we see that Δ is a known function of z in each layer of TZ. Particularly $-0.219 \leqq \Delta(z) \leqq -0.172$ if $220 \leqq z\,[\text{km}] \leqq 400$, $0.731 \leqq \Delta(z) \leqq 0.984$

if $400 \leqq z\,[\mathrm{km}] \leqq 600$, and $-0.633 \leqq \Delta(z) \leqq -0.627$ if $600 \leqq z\,[\mathrm{km}] \leqq 670$. Incidentally, the deviations of Δ from zero in TZ are on the whole about ten times greater than those of Δ in the lower mantle and core.

Using (1), the temperature gradient $\mathrm{d}T/\mathrm{d}z$ can be expressed by

$$\frac{\mathrm{d}T}{\mathrm{d}z} = g\left(\frac{\gamma T}{\Phi} - \frac{\Delta}{\gamma C_P}\right), \tag{3}$$

where C_P is the specific heat at constant pressure. There is no doubt that T is strictly increasing in the Earth's interior. Hence $\mathrm{d}T/\mathrm{d}z > 0$ at every z. The right member of equation (3) makes obvious that the term $g\gamma T/\Phi$ represents the so-called adiabatic temperature gradient at depth z if and only if $\Delta(z) = 0$ becomes true. However, this case does not happen in TZ. From (1) it follows that the signs of Δ and $\mathrm{d}S/\mathrm{d}z$ are unlike. As regards the slope $\mathrm{d}T/\mathrm{d}z$, it may unmistakably be called the superadiabatic (subadiabatic) temperature gradient everywhere $\Delta(z)$ takes on negative (positive) values.

We must not fail to mention that a neccessary condition for convective transport of the material at depth z is that $\mathrm{d}T/\mathrm{d}z$ is superadiabatic. Thus the medium would be out of convective equilibrium. From (2) we see that the negative values of Δ indicate a rate of compression less than isentropic as should be caused by thermal expansion from a conductive, "superadiabatic contribution" to $\mathrm{d}T/\mathrm{d}z$. On the other hand, having in view equation (2), the positive values of Δ suggest compression more rapid than that in a homogeneous medium under isentropic condition. Then it is quite conceivable that the depth range under consideration features phase changes [Brown and Shankland, 1981; Stacey, 1977].

The values of Δ are positive in the region intermediate between the upper and lower depth range of TZ where Δ takes on negative values. We now see why there is every reason to believe that phase transitions occur in the intermediate layer of TZ, as has already been emphasized. In this connection special interest must be paid to the state of the material located in a certain (infinitesimal) neighbourhood of either boundary of that layer. In either case $\Delta(z)$ undergoes a jump when crossing such a surface. Then $\Delta(z+0) - \Delta(z-0)$ amounts 0.903 or -1.618 if $z = 400$ km or $z = 600$ km, respectively.

Another feature of $\Delta(z)$ results in ranking as a minorant function of the always positive difference $C_P/C_V - 1$ at every z, where C_V is the specific heat at constant specific volume $V := \varrho^{-1}$. In evidence of that we make use of (3) and the relation $T = (C_V^{-1} - C_P^{-1})\,\Phi/\gamma^2$ [Ullmann, 1984b] to get

$$\frac{\mathrm{d}T}{\mathrm{d}z} = \frac{g}{\gamma C_P}\left(\frac{C_P}{C_V} - 1 - \Delta\right). \tag{4}$$

From this it immediately follows that $\mathrm{d}T/\mathrm{d}z > 0$ involves $\Delta < (C_P/C_V) - 1$. Since $\Delta > 0.731$ in TZ, this entails the absurd condition according to which C_P/C_V would be compelled to take on values greater than 1.7 in TZ.

Hence it appears that when applying familiar relations arising from equilibrium thermodynamics to the Earth's interior we run the risk of jumping at conclusions. Such relations are commonly used with the implicit understanding the substances spread over the whole depth range under consideration are homogeneous with respect to phase as well as chemical composition. But that is just what we do not expect of TZ. Therefore a different approach is needed to derive useful conclusions, even if only the intermediate layer of TZ is concerned.

On phase changes within TZ

Henceforth we shall confine ourselves to considering a one-component system whose chemical composition is uniform but in which two phases are present. That system at depth z suitably chosen is supposed to be subjected to work by hydrostatic pressure only. We further assume that immediately below the depth z, that is at $z + \varepsilon$ where $\varepsilon > 0$ is arbitrarily small, only the higher-pressure phase is located, whereas above z, that is at $z - \varepsilon$, the lower-pressure phase exists solely. Any quantity referring to one of the two phases shall be marked with the subscript 1 or alternatively 2. We ignore any possible surface effects at the interface between the phases, and therefore both pressure P and temperature T are uniform throughout the system as a whole.

The two phases of the substance at depth z are required to coexist in equilibrium. Evidently that system is kept at constant pressure, P, and temperature, T. The most convenient set of conditions sufficient to determine the equilibrium configuration is to minimize the specific Gibbs function G since its accompanying conditions, $\delta P = 0$ and $\delta T = 0$, are automatically fulfilled. We therefore make use of the relation $G = \mu_1 G_1 + \mu_2 G_2$, where G_i is the specific Gibbs function of the ith phase and μ_i is the respective mass fraction ($i = 1, 2$). Since we consider a closed system in which mass is conserved this is subject to the constraint $\mu_1 + \mu_2 = 1$. Hence, regarding $\delta\mu_1 = -\delta\mu_2$ and $(\delta G)_{P,T} = 0$, we find that G at depth z must have the same value in two phases in mutual equilibrium: $G = G_1(P, T) = G_2(P, T)$. Note that G_i is supposed to be continuous and differentiable as many times as required in some neighbourhood of the point P, T.

Proceeding on the assumption that $\partial(G_1 - G_2)/\partial P$ as well as $\partial(G_1 - G_2)/\partial T$ does not vanish at depth z we turn our attention to a first order phase transition occurring at that depth. In this case it is instructive to represent $G_i(P + \delta P, T + \delta T)$ as a surface in three dimensional space spanned by pressure, temperature and specific Gibbs function. The surface for the two different phases intersect in a line along which $G_1(P + \delta P, T + \delta T) = G_2(P + \delta P, T + \delta T)$ becomes true. Hence, for sufficiently small absolute values of δP and δT, it follows that approximately $[\delta P(\partial/\partial P)_T + \delta T(\partial/\partial T)_P]\,(G_1 - G_2) = 0$ or, since $(\partial G_i/\partial P)_T = V_i$ and $(\partial G_i/\partial T)_P = -S_i$, $\delta P/\delta T = (S_2 - S_1)/(V_2 - V_1)$, provided $V_1 \neq V_2$ and $S_1 \neq S_2$ as required above.

It is expedient to replace V_i and S_i by the respective one-sided limits $V(z \pm 0) := \lim_{\varepsilon \to 0} V(z \pm \varepsilon)$ and $S(z \pm 0) := \lim_{\varepsilon \to 0} S(z \pm \varepsilon)$. We may then write

$$\frac{\delta P}{\delta T} = \frac{S(z + 0) - S(z - 0)}{V(z + 0) - V(z - 0)}. \tag{5}$$

This is the Clausius-Clapeyron equation which gives the rate at which the pressure would have to change with temperature for the two phases at depth z to remain in equilibrium. In other words, formula (5) yields the gradient of the phase boundary in the pressure-temperature plane at point P, T. Since the density is strictly increasing at every depth $z > 220$ km, it particularly follows that $V(z+0) - V(z-0) < 0$. Therefore $\delta P/\delta T$ and $S(z+0) - S(z-0)$ have unlike signs.

Let G_1 and S_1 (G_2 and S_2) be related to the lower-pressure (higher-pressure) phase so that G is identical with G_1 or G_2 at every depth $z - \varepsilon$ or $z + \varepsilon$, respectively. Then, since the temperature is strictly increasing with depth throughout the Earth, and G takes on a minimum at depth z, we decide that $(\partial G_1/\partial T)_P = -S(z-0)$ must be greater than $(\partial G_2/\partial T)_P = -S(z+0)$. The positive expression $T[S(z+0) - S(z-0)]$ represents the specific heat transformation due to the first order phase change occurring at depth z. Thus the slope $\delta P/\delta T$ resulting from (5) is found to take on a negative value.

According to PREM the density undergoes a jump at $z = 400$ km and nowhere else in TZ. The corresponding anomaly in density given by $\varrho(z+0) - \varrho(z-0)$ with $z = 400$ km amounts about $0.18 \cdot 10^3$ kg/m³. It is taken for granted that the comparatively sharp interface at that depth is attributed to a first order phase transition. RINGWOOD [1982] pointed out once again that the seismologically observed discontinuity at $z = 400$ km is due to the transformation of olivine to $\beta(Mg,Fe)_2SiO_4$ [JEANLOZ and THOMPSON, 1983; LEES, BUKOWINSKI and JEANLOZ, 1983].

It is evident that a phase change may occur at depth z even though both density and specific entropy are continuous there. Then the representations $V(z \pm \varepsilon) = V(z) \pm V'(z \pm 0)\,\varepsilon + 0(\varepsilon)$ and, correspondingly, $S(z \pm \varepsilon) = S(z) \pm S'(z \pm 0)\,\varepsilon + 0(\varepsilon)$ if $\varepsilon \to 0$ prove suitable. Here and hereafter we do not think it necessary to explain $V'(z \pm 0)$ and $S'(z \pm 0)$ as well as further notations of that kind, formula (5) thus yields $\delta P/\delta T = [S'(z+0) - S'(z-0)]/[V'(z+0) - V'(z-0)]$.

Owing to $V := \varrho^{-1}$ we now have $V'(z \pm 0) = -\varrho'(z \pm 0)/\varrho(z)^2$. The defining equation (1) gives rise to the relations $S'(z \pm 0) = -[g(z)/T(z)]\,[\Delta(z \pm 0)/\gamma(z \pm 0)]$ in which the property of both temperature and acceleration of gravity of being a continuous function of depth throughout the Earth is taken into account. Therefore the slope $\delta P/\delta T$ arising from a phase transition of order higher than first which occurs at depth z can be expressed by

$$\frac{\delta P}{\delta \ln T} = \left[\frac{\Delta(z+0)}{\gamma(z+0)} + \frac{\Delta(z-0)}{\gamma(z-0)}\right]\left[\frac{g(z)\,\varrho(z)^2}{\varrho'(z+0) + \varrho'(z-0)}\right]. \tag{6}$$

This leads with the aid of (2) and the one-sided limits of the effective incompressibility given by $\varkappa(z \pm 0) = g(z)\,\varrho(z)^2/\varrho'(z \pm 0)$ to

$$\frac{\delta P}{\delta \ln T} = \left[\frac{\Delta(z+0)}{\gamma(z+0)} + \frac{\Delta(z-0)}{\gamma(z-0)}\right]\left[\frac{\Delta(z+0)+1}{\varkappa_S(z+0)} + \frac{\Delta(z-0)+1}{\varkappa_S(z-0)}\right]^{-1}. \tag{7}$$

Since ϱ, v_p and v_s are continuous at every depth in TZ except $z = 400$ km, we have $\varkappa_S(z \pm 0) = \varkappa_S(z)$ throughout TZ, apart from $z = 400$ km. Thus (7) reduces to

$$\frac{\delta P}{\delta \ln T} = \varkappa_S(z)\left[\frac{\Delta(z+0)}{\gamma(z+0)} + \frac{\Delta(z-0)}{\gamma(z-0)}\right][\Delta(z+0) + \Delta(z-0) + 2]^{-1}. \tag{8}$$

Ehrenfest was the first who introduced a classification scheme whereby the order of a phase transition is defined as the order of the lowest differential of the (specific) Gibbs function which shows a discontinuity at the transition. Adopting his concept we can only state that formulas (6)—(8) require a phase change at depth z which is of order higher than first. However, considering the fact that a transition of order higher than second has not yet been observed we henceforth assume that the slope $\delta P/\delta T$ according to (6)—(8) must be ascribed to a second order phase change.

The important thing about it is that the two field functions Φ and γ cannot be continuous at depth z simultaneously if a transition of that type occurs there. To verify this statement we recall that, by definition, $V_1(P + \delta P, T + \delta T) = V_2(P + \delta P, T + \delta T)$ must than come true. Hence, for sufficiently small absolute values of δP and δT, $[\delta P(\partial/\partial P)_T + \delta T(\partial/\partial T)_P]\,(V_1 - V_2) = 0$. This yields

$$\frac{\delta P}{\delta T} = \frac{\alpha(z+0) - \alpha(z-0)}{\varkappa_T(z+0)^{-1} - \varkappa_T(z-0)^{-1}}, \tag{9}$$

where $\varkappa_T := -V(\partial P/\partial V)_T$ is the isothermal incompressibility and $\alpha := \varrho(\partial V/\partial T)_P$ is the isobaric cubic expansivity. Equation (9) is known as the second Ehrenfest relation.

The familiar thermodynamic formula $\varkappa_S/\varkappa_T = 1 + T\gamma\alpha$ renders possible te replace the denominator on the right-hand side of (9) by $\varkappa_S(z+0)^{-1}\,[1 + T(z)\,\gamma(z+0)\,\alpha(z+0)] - \varkappa_S(z-0)^{-1}\,[1 + T(z)\,\gamma(z-0)\,\alpha(z-0)]$. From this we learn that if $\varkappa_S(z \pm 0) = \varkappa_S(z)$ as well as $\gamma(z \pm 0) = \gamma(z)$ would be correct then equation (9) degenerates into $\delta P/\delta \ln T = \varkappa_S(z)/\gamma(z)$. In contrast to this, equation (9) gives then rise to $\delta P/\delta \ln T = [\varkappa_S(z)/\gamma(z)]/\{1 + 2/[\Delta(z+0) + \Delta(z-0)]\}$ which is obviously in contradiction to the previous expression since Δ^{-1} differs from zero considerably. Persisting in conformity with PREM we therefore get

$$\frac{\delta P}{\delta \ln T} = \varkappa_S(z)\,\frac{\alpha(z+0) - \alpha(z-0)}{\gamma(z+0)\,\alpha(z+0) - \gamma(z-0)\,\alpha(z-0)}. \tag{10}$$

as a parallel to (8), minding that either relation refers to TZ.

So far we are short of detailed information on the field functions γ and α, in particular whenever second order phase transitions must be taken into consideration. For this reason alone the last formula is unsuitable for both computation and sign determination of the slope $\delta P/\delta T$. To find out the latter we are dependent on equations (6)—(8). From them we get

$$\operatorname{sgn}\frac{\delta P}{\delta T} = \operatorname{sgn}\,[\Delta(z+0)\,\gamma(z-0) + \Delta(z-0)\,\gamma(z+0)]. \tag{11}$$

We see that if $\Delta(z+0)$ as well as $\Delta(z-0)$ is positive then $\delta P/\delta T$ takes on a positive value.

According to PREM, we must first take the view that the only discontinuity surface in TZ possibly due to a second order phase change is located at $z = 600$ km. Making reference to that depth we have $\Delta(z+0)/\Delta(z-0) = -0.644$, and therefore $\operatorname{sgn}\delta P/\delta T = \operatorname{sgn}\,[\gamma(z+0)/\gamma(z-0) - 0.644]$ at $z = 600$ km. There is reason to suppose that the ratio $\gamma(z+0)/\gamma(z-0)$ formed at every depth z in TZ does not differ from unity such that it takes on a value less than 0.644. As a result, the slope $\delta P/\delta T$ appertaining to the phase transition possibly occurring at $z = 600$ km must be positive.

In search of further phase transitions taking place within TZ we find out by PREM only that, if at all, the depth range $400 < z\,[\text{km}] < 600$ is suitable for the occurrence of second order phase changes. From (8) we then have, at least approximately, $\delta P/\delta \ln T = [\varkappa_S(z)/2]\,[\gamma(z+0)^{-1} + \gamma(z-0)^{-1}]/[1 + \Delta(z)^{-1}]$ or $\delta \ln P/\delta \ln T \approx 3[\gamma(z+0)^{-1} + \gamma(z-0)^{-1}]$, $400 < z\,[\text{km}] < 600$. Note that $\delta P/\delta T$ is positive throughout that open depth interval.

In conclusion we revert to the interpretation of a phase change. Now that a second order phase transition is considered, the two phase system cannot pass from one of the G_i surfaces to another. The absence of transformation heat or density change would require the gradients to be identical at the transition, and the jumps of the second order differential coefficients of G_i would require one curvature to be greater than the other. But then the surfaces would touch but not cross at the transition.

We are therefore forced to the conclusion that one and the same G surface is involved in a second order transition and that some property of the two phase system causes the second partial derivatives of G to change discontinously. This conclusion is supported by the absence of metastable states in second and higher order transitions. Superheating and supercooling have never been found in transitions other than first order, so that only there is any evidence of the continuity of G for a particular phase through the transition [Adkins 1975].

Conclusions

The parameter $\Delta(z)$, originally defined by (1), becomes computable at every depth z only with the aid of data originating from PREM by virtue of the transformation into (2). In consequence of PREM the transition zone, TZ, consists of three consecutive depth intervals reaching from $z = 220$ km down to $z = 670$ km. The values of $\Delta(z)$ are positive for the intermediate interval $400 \leqq z\,[\text{km}] \leqq 600$, whereas they are negative for the two outer subintervals of TZ. From (2) we also learn that $\Delta(z)$ proves tantamount to the relative deviation of the effective compressibility, $\varkappa^{-1}$, from that which would arise from isentropy at z, that is $\varkappa_S{}^{-1}$.

Hence, in accordance with the relations (3) and (4), it follows that the negative values of Δ in the upper as well as the lower subregion of TZ should be caused by thermal expansion from a conductive, superadiabatic contribution to the temperature gradient dT/dz needed to maintain convective motion, while the large values of Δ in the intermediate layer of TZ indicate a subadiabatic temperature gradient dT/dz due to a greater-than-adiabatic compression suggestive of phase changes. Since it is beyond doubt that the temperature $T(z)$ is strictly increasing at every depth z we gather from (4) that the ratio C_P/C_V of the two principal specific heats C_P and C_V must be greater than $\Delta + 1$. Consequently the particularly great values of Δ for the interior of TZ involve a lower bound of C_P/C_V at about 1.7. This bound, however, is too high to be consistent with the common working hypothesis which implies a solid substance homogeneous with respect to phase as well as chemical composition.

Proceeding instead on the assumption that phase transitions occur within TZ we think it best to abide by considering one-component systems. This makes it possible to examine the details of PREM for information about phase changes taking place in TZ. Attention is confined to the parameter of particular importance given by the slope $\delta P/\delta T$. It represents the rate at which the pressure P would have to change with temperature for the two phases adjacent at depth z to remain in equilibrium.

According to PREM each of the seismic velocities $v_p(z)$ and $v_s(z)$ as well as the density $\varrho(z)$ undergoes a jump at $z = 400$ km but nowhere else within TZ. Hence this seismologically discovered discontinuity is attributed to a first order phase transition. The relevant Clausius-Clapeyron slope $\delta P/\delta T$ proves to be negative. Its absolute value cannot be estimated as long as there is a lack of information on the corresponding jump quantity of the specific entropy. From geochemical considerations it follows that the comparatively sharp discontinuity at $z = 400$ km may be due to the phase transformation of olivine to the modification $\beta(Mg, Fe)_2SiO_4$.

With regard to the fact that a phase change of order higher than second after Ehrenfest has not yet been observed we also infer from PREM that all further transitions occurring with TZ must be of second order. In that case, since each of the field functions $v_p(z)$, $v_s(z)$ and $\varrho(z)$ is continuous in TZ except for $z = 400$ km, we are bound to make allowance for a jump of the Grüneisen parameter $\gamma(z)$ at the depth in question. Again, unaware of that jump quantity we are unable to estimate the absolute value of the slope $\delta P/\delta T$ for a second order phase change. In a recent paper [ULLMANN, 1985] attention was not paid to this special feature.

In search of such a phase transition occurring within TZ we are referred by PREM to the depth $z = 600$ km at which the gradients of $v_p(z)$, $v_s(z)$ and $\varrho(z)$ are all discontinuous. Considering the available information about γ we have no doubt that the sign relation (11) ensures the slope $\delta P/\delta T$ to be positive at that depth.

Hence PREM gives rise to the assumption that there exists a first order phase change with negative slope $\delta P/\delta T$ at $z = 400$ km as well as a second order transition with positive $\delta P/\delta T$ at $z = 600$ km. Moreover, any further phase change to be expected in TZ must be intermediate between these depths. It must apparently be of second order so that its slope $\delta P/\delta T$ takes on a positive value because Δ is greater than zero throughout that region.

Of course, in taking the PREM data too literally we are liable to overestimate the actual information content of our study. The mere fact that most probably the transient behaviour of a geophysical field function cannot be related to a perfectly flat discontinuity at depth z warns against rushing to conclusions. Also, the smoothing by PREM precludes from exhausting the occurrence of phase transitions in TZ. Nevertheless, the ascertainment that every discontinuity within TZ below $z = 400$ km is caused by a structural transformation of second order with positive slope $\delta P/\delta T$ is worthy of notice.

The values of Δ in TZ should be correct on the whole. Therefore it is also worthwile to pursue the implications of the use of this parameter, if only as indication of what can be gained from more refined seismic observations at a later time. Such explorations must be aimed at discovering comparatively slight discontinuities of the two seismic velocity gradients which are attended by a small jump in density. So RINGWOOD [1982] pointed out that further transformation of the above-mentioned β phase to $(Mg, Fe)_2SiO_4$ (spinell) may cause a weak seismic discontinuity near $z = 550$ km [NIAZI and ANDERSON, 1965; JOHNSON, 1967; ARCHAMBEAU, FLINN and LAMBERT, 1969; KING and CALCAGNILE, 1976; ENGLAND, WORTHINGTON and KING, 1977; FUKAO, 1977; BURDICK and HELMBERGER, 1978; GIVEN and HELMBERGER, 1980; BURDICK, 1981; FUKAO, NAGAHASHI and MORI, 1982; KLINGE, 1984].

What the PREM data actually show is that the slope $\delta P/\delta T$ has opposite signs for the dominant phase transitions occurring at $z = 400$ km and $z = 600$ km. If we consider the down-going material in mantle convection to be cooler than average and, conversely, the up-going material to be hotter than the depth at which a particular phase change happens in each of these can be determined by $\delta P/\delta T$. The elevations and depressions of the two principal phase transitions taking place within TZ prove to be negatively correlated. Evidence that this is in fact so is presented in the analysis of global gravity data by McQueen and Stacey [1976; cf. also Dorman and Lewis, 1971].

References

Adkins, C. J.: Equilibrium thermodynamics. McGraw Hill, London (1975).

Archambeau, C. B.; Flinn, E. A.; Lambert, D. G.: Fine structure of the upper mantle. J. geophys. Res. **74** (1969) 5825–5865.

Brown, J. M.; Shankland, T. J.: Thermodynamic parameters in the Earth as determined from seismic profiles. Geophys. J. R. astr. Soc. **66** (1981) 579–596.

Burdick, L. J.: A comparison of upper mantle structure beneath North America and Europe, J. geophys. Res., **86** (1981) 5926–5936.

Burdick, L. J.; Helmberger, D. V.: The upper mantle P-velocity structure of the western United States. J. geophys. Res. **83** (1978) 1699–1712.

Dorman, L. M.; Lewis, B. T.: Evidence supporting a dipolar compensation mechanism. EOS, Trans. Am. geophys. Un. **52** (1971) 353 (abstract).

Dziewonski, A. M.; Anderson, D. L.: Preliminary reference Earth model. Phys. Earth planet. Int. **25** (1981) 297–365.

England, P. C.; Worthington, M. H.; King, D. W.: Lateral variations in the structure of the upper mantle beneath Eurasia. Geophys. J. R. astr. Soc. **48** (1977) 71–79.

Fukao, Y.: Upper mantle *P*-structure and the 650-km discontinuity. In: High Pressure Research. Eds. Manghnani, M. H.; Akimoto, S., Academic Press, New York 1977.

Fukao, Y.; Nagahashi, T.; Mori, S.: Shear velocity in the mantle transition zone. In: High Pressure Research in Geophysics. Eds. Akimoto, S.; Manghnani, M., Academic Publish. Japan, Tokyo 1982.

Given, J. W.; Helmberger, D. V.: Upper mantle structure of northwestern Eurasia. J. geophys. Res. **85** (1980) 7183–7194.

Jeanloz, R.; Thompson, A. B.: Phase transitions and mantle discontinuities. Rev. Geophys. Space Phys. **21** (1983) 51–74.

Johnson, L. R.: Array measurements of P-velocities in the upper mantle. J. geophys. Res. **72** (1967) 6309–6325.

King, D. W.; Calcagnile, G.: P-wave velocities in the upper mantle beneath Fennoscandia and western Russia. Geophys. J. R. astr. Soc. **46** (1976) 407–432.

Klinge, K. D.: Discontinuities in the Earth's crust and mantle discovered by seismogram stacking. Gerlands Beitr. Geophys., **93** (1984) 323–325 (in German).

Lees, A. C.; Bukowinski, M. S. T.; Jeanloz, R.: Reflection properties of phase transition and Mccompositional change models of the 670-km discontinuity. J. geophys. Res., **88** (1983) 8145 to 8159.

Queen, H. W. S.; Stacey, F. D.: Interpretation of l ow degree components of gravitational potential in terms of undulations of mantle phase boundaries. Tectonophys. **34** (1976) T1–T8.

Niazi, M.; Anderson, D. L.: Upper mantle structure of western North America from apparent velocities of P-waves. J. geophys. Res. **70** (1965) 4633–4640.

Ringwood, A. E.: Phase transformations and differentiation in subduced litosphere: implications for mantle dynamics, basalt petrogenesis, and crustal evolution. J. Geol. **90** (1982) 611 to 643.

Stacey, F. D.: Applications of thermodynamics to fundamental Earth physics. Geophys. Surv. **3** (1977) 175–204.

Ullmann, W.: On the parametrization of inhomogeneities in the Earth. Gerlands Beitr. Geophys. **93** (1984a) 59–68.

Ullmann, W.: Boundary condition for the inhomogeneity parameter in the Earth: implications for heat action within the upper mantle. Gerlands Beitr. Geophys. **93** (1984b) 142–148.

Ullmann, W.: On thermal features within the transition zone. Phys. Earth planet. Int. (1985) in press.

Effects of high pressure phase transitions on chemical differentiation

S. Franck, H. Stiller, W. Seifert

Introduction

The Earth's mantle, which represents about 2/3 of the Earth's mass, is usually considered to be mostly homogeneous in composition. This picture is also the basic assumption in many investigations of mantle convection. Discontinuities in the mantle are often interpreted as isochemical phase changes in an olivine-rich material rather than chemical boundaries. In this way the 400 km and 670 km seismic discontinuities are generally assumed to be the olivine-spinel and spinel-post-spinel phase boundaries (see e.g. Ringwood, 1975; Zharkov and Trubitsyn, 1978).

In the present paper we investigate a physical mechanism that may cause a chemical differentiation at a polymorphic phase transition. The idea is based on the simple picture that during the density increase at the transition there is the appearance of strong stress-fields that act on the constituents of the solid material. Particularly so-alled incompatible ions with ionic radii not appropriate for the high-pressure phase may drift to the grain boundaryaand reach high-diffusivity paths so that in this way they may be enriched in the low-pressure phase (see Figures 1 and 2). It is important that our proposed mechanism of differentiation is not related to partial melting as in the usual models in geochemistry (see e.g. Yoder, 1976) but acts in a solid. Therefore this mechanism may be very favoured at depths greater than 200 km where partial melting seems not easily possible as follows from investigations on the temperature distribution and the melting curve in the Earth's mantle (see e.g. Green and Ringwood, 1970). Numercial estimations show the necessity that the solid is in a superplastic state at the phase transition as already discussed by Kalinin and Rodkin (1982) in connection with the earthquake mechanism and by Parmentier (1981) and Poirier (1982) in connection with other phenomena in the mantle.

In the next section we describe the physical model and give numerical estimations for the 400 km and 670 km seismic discontinuities. In the last section we discuss geochemical implications.

Physical model and calculations

Changes of concentrations during phase transitions in multicomponent systems may be discribed with the help of the usual diffusion equation.

$$\frac{\partial C}{\partial t} = \nabla(D\nabla C), \tag{1}$$

where $C = C(x, y, z, t)$ is the concentration and D the diffusion coefficient.

As already mentioned above, the increase in density at a pressure-induced phase transition produces a strong stressfield and the corresponding interaction energy V of an ion with radius r is (Aleksandrov and Ljubov, 1963):

$$V = -\frac{4}{3}\,\pi r^3 \sigma_{ii}. \tag{2}$$

In this formula σ_{ii} is the sum of the diagonal elements of the stress tensor.

This is a simple mechanical model of a complicated process that can include also chemical effects which are not considered here. Nevertheless, we think that equation (2) contains the main part of the whole potential.

The sum of diagonal elements of the stress tensor that may be related to bulk modulus K and strain u_{ii} or density jump (Landau and Lifschitz, 1965; Kittel, 1973):

$$\sigma_{ii} = -3K\,\frac{\varrho_2 - \varrho_1}{\varrho_2}. \tag{3}$$

The elastic potential V causes a flux of material that is directed from the high-pressure to the low-pressure phase. Therefore the diffusion equation takes the following form (K_B is Boltzmann's constant, T temperature):

$$\frac{\partial C}{\partial t} = D\Delta C + \frac{DC}{K_B T}\,\Delta V + \frac{D}{K_B T}\,\nabla C \nabla V. \tag{4}$$

The first term of the right-hand side stands for usual diffusion, the second for stress-induced drift and the third is a mixed term of both diffusion and drift.

In the case of a strong stress-field we may neglect the first term in comparison with the others (Aleksandrov and Ljubov, 1963). In our investigation we have to consider the diffusion equation together with the equation of continuity for the flux at the phase boundary:

$$+\,(C_b - C_2)\,v = -\frac{DC}{K_B T}\,\nabla V|_b. \tag{5}$$

The indices 1 (see e.g. equation (7)), 2 and b refer always to the low-pressure phase, high-pressure phase and phase boundary, respectively. The velocity of propagation of the phase boundary is v. Equation (5) describes the flux at the phase boundary within the grain. Even in the case of multivariate phase transformations of mantle material where the transition region as a finite thickness of a few kilometers (see Fig. 1), each individual grain has to undergo the transformation and we may apply the picture as shown in Fig. 2.

Equation (4), without the first term of the right-hand side, and equation (5) were

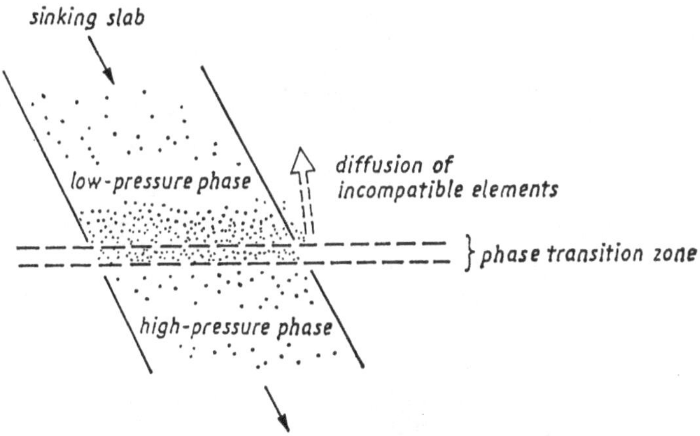

Fig. 1. Macroscopic view of the proposed effect of chemical differentiation at phase boundaries in sinking slabs.

solved with the help of a dimensionless variable u in a moving coordinate system:

$$u = \frac{x}{R} - \frac{vt}{R} + 1, \tag{6}$$

where x is the coordinate of phase boundary propagation and R is the characteristic dimension of a grain (see Fig. 3). The new variable u has the following interesting properties:

$u > 1$ in the low-pressure phase (lpp)

$u < 1$ in the high-pressure phase (hpp)

$u = 1$ at the phase boundary (b).

As the result of this transformation (see appendix) we arrive at the following equations:

$$C_b = C_1 \left(\frac{-\frac{\pi}{3} r^3 \sigma_{ii} - \frac{Rv}{D} K_B T}{-\frac{4\pi}{3} r^3 \sigma_{ii} - \frac{Rv}{D} K_B T} \right), \tag{7}$$

$$(C_b - C_2) \frac{Rv}{D} K_B T = \frac{-4\pi}{3} r^3 \sigma_{ii} C_b . \tag{8}$$

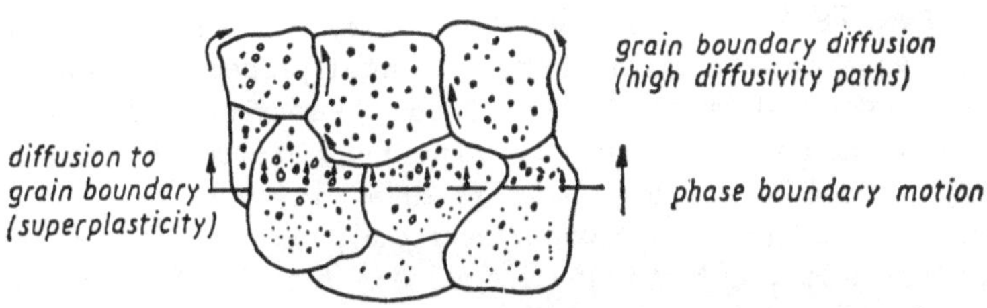

Fig. 2. Microscopic view of the drift of incompatible ions to the grain boundary as the rate-limiting step of ion transport and further movement on high-diffusivity paths at grain boundaries.

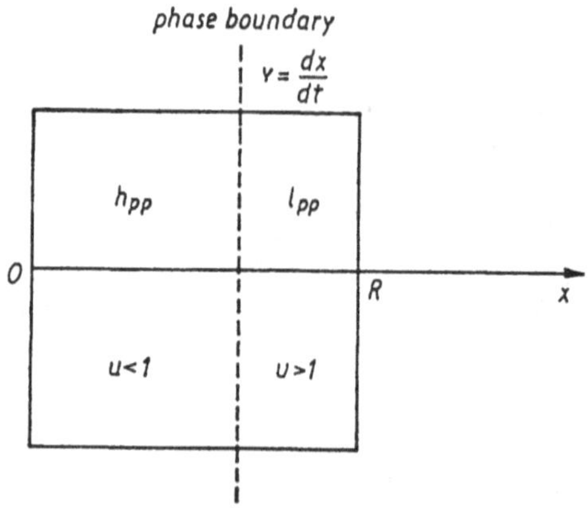

Fig. 3. Moving coordinate system for the description of phase boundary motion.

The combination of equations (3), (7) and (8) (see appendix) gives the following formula for the change of concentrations of incompatible ions with radius r in a grain of characteristic dimension R:

$$\frac{C_2}{C_1} = 1 - \frac{\pi K \frac{\varrho_2 - \varrho_1}{\varrho_2} D}{R v K_B T} r^3. \tag{9}$$

This is a steady state solution that attains much longer than the characteristic time for a convective motion in the mantle. Formula (9) was analysed for a sinking slab crossing the 400 km and 670 km phase boundaries. The most problematic quantity is the diffusion coefficient D for the corresponding thermodynamical conditions. For the purpose of using a characteristic interpolated value for cation diffusion we apply the well-known formula

$$D = D_0 \exp\left[-(E^* + pV^*)/K_B T\right] \tag{10}$$

(LEES et al., 1983; STOCKER and ASHBY, 1973), where D_0 is the high-temperature diffusion coefficient, E^* is the activation energy, and V^* is the activation volume.

Applying current numerical values, as given in Table 1, we find no effect for all relevant ionic radii, even with help of the relatively high estimate of the temperature at 670 km of LEES et al. (1983). But we should remember that at a phase boundary there may be some differences in thermodynamical properties compared with the stable phase (STILLER and FRANCK, 1980). The local variation of D, p, V^*, E^*, and T in the region near the phase boundary was not taken into account because in this case an analytical solution of the problem is not possible.

One example of the specific behaviour near the phase boundary is the softening of the bulk modulus leading to an effective value of $K \simeq 50$ GPa (TRUBITSYN, 1979). This is of course different from the equilibrium value, but should be used in our calculation. As discussed by KALININ and RODKIN (1982), there have been observed

Table 1. Parameter estimates

Parameter/quantity	symbol	value at 400 km boundary	value at 670 km boundary	remarks
bulk modulus	K	50 GPa	50 GPa	estimation from TRUBITSYN (1979)
density of high-pressure phase	ϱ_2	3.8 g/cm³	4.4 g/cm³	from TRUBITSYN (1979)
density of low-pressure phase	ϱ_1	3.6 g/cm³	4.1 g/cm³	from TRUBITSYN (1979)
characteristic dimension of grains	R	1.5 cm	1.5 cm	from POIRIER (1981)
velocity of phase boundary propagation	v	8 cm/a $= 2.6 \cdot 10^{-7}$ cm/s	8 cm/a $= 2.6 \cdot 10^{-7}$ cm/s	subduction velocity, see POIRIER (1981)
pressure	p	13 GPa	24 GPa	from TRUBITSYN (1979)
Temperature	T	1880 K	2300 K 2500 K	from TRUBITSYN (1979) from LEES et al. (1983)
high-temperature diffusion coefficient	D_0	$1.2 \cdot 10^6$ cm²/s	$1.2 \cdot 10^6$ cm²/s	from LEES et al. (1983)
activation energy	E^*	644 kJ/mol	644 J/mol	from LEES et al. (1983)
superplastic activation energy	$\frac{1}{2} E^*$	322 kJ/mol	322 kJ/mol	see text
activation volume	V^*	11 cm³/mol	11 cm³/mol	from LEES et al. (1983)
radii of incompatible ions	r	$0.5 \ldots 2.0 \cdot 10^{-8}$ cm	$0.5 \ldots 2.0 \cdot 10^{-8}$ cm	see Fig. 3
ratio of concentrations in high- and low-pressure phase	C_2/C_1	1 $1 - 0.1 \cdot 10^{24}\,\mathrm{cm}^{-3}\,\mathrm{r}$	1 $1 - 0.05 \cdot 10^{24}\,\mathrm{cm}^{-3}\,\mathrm{r}$ $1 - 0.54 \cdot 10^{24}\,\mathrm{cm}^{-3}\,\mathrm{r}$	with usual activation Energy E superplasticity, $T = 2300$ K at 670 km superplasticity, $T = 2500$ K at 670 km

anomalously high velocities of diffusion (some orders of magnitude greater) at phase transitions that cause the phenomenon of superplasticity.

As described by POIRIER (1976), a superplastic state is characterized by an activation energy in the order of the activation energy for grain boundary diffusion which on the other hand is about one half of the usual activation energy for bulk diffusion. Therefore we also present numerical estimations for the superplastic state by using $1/2E^*$ as activation energy. As can be seen from Table 1 and Fig. 4, in this case there is a noticeable effect of a chemical differentiation at both phase boundaries on the Earth's mantle.

Even if parameters like grain size, velocity of phase boundary motion, bulk modulus, and density jump are slightly changed there is alsways the effect of a chemical differentiation if superplasticity is taken into account.

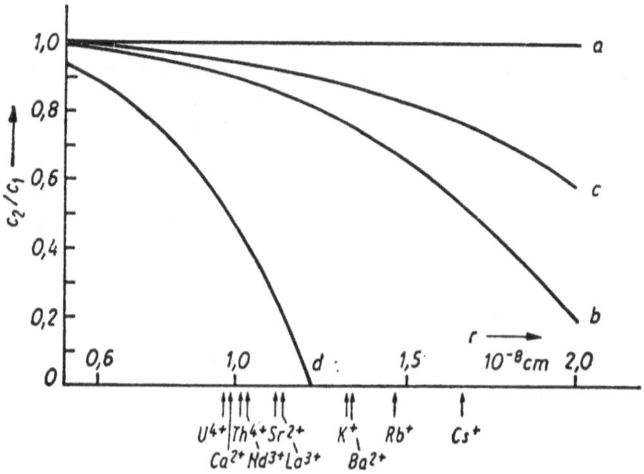

Fig. 4. Ratios of ion concentrations in low- and high-pressure phase.
a: both phase boundaries with usual activation energy
b: at 400 km boundary with superplasticity
c: at 670 km boundary with superplasticity and $T = 2300$ K
d: at 670 km boundary with superplasticity and $T = 2500$ K

Conclusions

Our proposed mechanism of a chemical differentiation at phase boundaries gives reasonable numerical values for the enrichment of incompatible ions above the 400 km discontinuity. In the case of olivine-spinel transition this may be particularly interesting for the vertical differentiation of Ca^{2+}, Al^{3+}, and Ti^{4+}. Assuming similar relations for the phase transition of the pyroxene-garnet component at about 400 km, we find effects for many incompatible ions. In the case of K^+, Br^{2+}, and Ba^{2+} we may also give a geochemical interpretation of the high content of these ions in alkali basalts kimberlites, and carbonatites that are believed to come from great depths of the upper mantle via diapirs. Taking C_2/C_1 0.4 as a typical ratio for the 400 km discontinuity and using results of POIRIER (1981) and LELIWA-KOPYSTINSKI and TEISSEYRE (1984) for the thickness of the phase transition zone, we find concentration gradients of about 0.005...0.03 km in this region. Such concentration gradients are in the same order of magnitude as in the case of partial melting. The application of our model to the 670 km discontinuity is related to the problem of whole mantle convection. But according to current thiking in geochemistry (ALLEGRE et al., 1983) and planetary cosmogony (STILLER et al., 1984) there has been at least an early phase in the Eath's history when convection of the mantle as a whole occured. We should remember that the ratio of concentrations is changed during every cycle of convection, so that there could have been an important chemical differentiation even in early stages of the evolution. It is evident that these proposed effects of a chemical differentiation should be taken into account in isotope analyses of mantle-derived rocks. Finally we want to refer to modern work about the problem of a compositional change at the 670 km discontinuity (LEES et al., 1983). According to this work, there may be a change in the iron/silicon ratio at this boundary. Our model was formulated for the description of the behaviour of foreign ions during the phase transition of the host crystal. But if the phase transition in the host crystal is diffusion-controlled (SUNG and BURNS, 1978) there may be the possibility of using similar considerations also for a quantitative calculation of iron/silicon ratios.

Appendix: numerical calculations

Our starting point is equation (4). In the case of a strong stress-field we may neglect the first term (pure diffusion) in comparison with the drift and the mixed term.

$$\frac{\partial C}{\partial t} = \frac{D}{K_B T} C \Delta V + \frac{D}{K_B T} \nabla C \nabla V . \tag{A1}$$

Equation (A1) is solved with the help of a new variable u in a moving coordinate system:

$$u = \frac{x}{R} - \frac{vt}{R} + 1,$$

$$C(x, t) = \Phi(u),$$

$$V(x, t) = F(u).$$

This leads to the following expressions:

$$\frac{\partial C}{\partial t} = (-)\, \Phi' \cdot \frac{v}{R},$$

$$\nabla C = \Phi' \cdot \frac{1}{R},$$

$$\nabla V = F' \cdot \frac{1}{R},$$

$$\Delta V = F'' \frac{1}{R^2}.$$

Inserting in (A1) gives a differential equation depending only on variable u:

$$-\Phi' \frac{v}{R} = \frac{D}{K_B T} \Phi F'' \frac{1}{R^2} + \frac{D}{K_B T} \Phi' \frac{1}{R} \cdot F' \frac{1}{R}.$$

With the abbreviation $p = RvK_B T/D$ we arrive at

$$-\Phi' \cdot p = \Phi F'' + \Phi' F' . \tag{A2}$$

In the same way as ALEKSANDROV and LJUBOV (1963) we use the transformed interaction energy in the form

$$F(u) = -\frac{4}{3} \pi r^3 \sigma_{ii}/u = A/u . \tag{A3}$$

At the phase boundary ($u = 1$) this expression is identical to the interaction energy (2). But in the lpp ($u > 1$) and hpp ($u < 1$), respectively, there is such a scaling that the gradient F' favours an accumulation of incompatible ions in the lpp. Putting (A3) in (A2) gives

$$\Phi'\left(-p\frac{u^3}{2A}+\frac{u}{2}\right)=\Phi. \tag{A4}$$

Equation (A4) can be integrated:

$$\int\limits_{\Phi_1}^{\Phi}\frac{d\Phi}{\Phi}=\int\limits_{2}^{u}\frac{du}{\left(-p\frac{u^3}{2A}+\frac{u}{2}\right)}. \tag{A5}$$

In (A5) Φ_1 is the undisturbed concentration of the lpp ($u=2$; $x=R$, $t=0$).

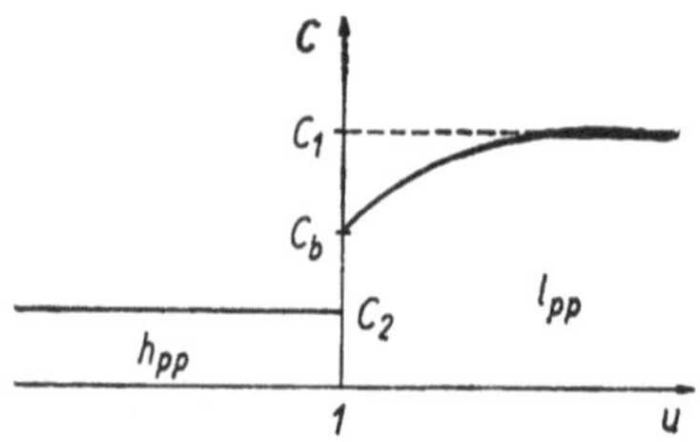

Fig. 5. Distribution of concentrations at the phase boundary.

(A5) can be solved with the help of usual integral tables, e.g. Bronstein and Semendjajew (1969) giving

$$\ln\left(\frac{\Phi}{\Phi_1}\right)=\ln\frac{u^2}{-u^2+A/p}\Bigg|_2^u, \tag{A6}$$

$$\frac{\Phi}{\Phi_1}=\frac{A/4-p}{A/u^2-p}. \tag{A7}$$

If we take (A7) at the phase boundary ($\Phi=\Phi_b$, $u=1$) and return to the original functions and variables we arrive exactly at equation (7) of the text.

Now we turn to the equation of continuity for the flux at the phase boundary (formula (5) of the text). The situation is sketched in Fig. 5.

$$(C_b-C_2)\,v=-\frac{DC}{K_BT}\nabla V\Bigg|_b. \tag{A8}$$

Formula (A8) is transformed with the same abbreviations and definitions as (A1) leading to

$$(\Phi_b-\Phi_2)\,p=A\cdot\Phi_b. \tag{A9}$$

Using (A7) at the phase boundary ($\Phi=\Phi_b$, $u=1$) and inserting in (A9) leads to the following expression:

$$\frac{\Phi_2}{\Phi_1}=1-\frac{A}{4p}. \tag{A10}$$

Returning to original functions and variables equation (A10) is exactly formula (9) of the text.

References

ALEKSANDROV, L. N.; LJUBOV, B. YA.: On the theory of growth of bainite crystals during the inter-mediate transition of austenite. Dokl. Akad. Nauk SSSR, **151** (1963) 3, 552–555.

ALLEGRE, C. J.; HART, S. R.; MINSTER, J. F.: Chemical structure and evolution of the mantle and continents determined by inversion of Nd and Sr isotopic data, I. Theoretical methods. Earth Planet. Sci. Lett., **66** (1983) 177–190.

BRONSTEIN, I. N.; SEMENDJAJEW, K. A.: Taschenbuch der Mathematik. BSB B. G. Teubner Verlagsgesellschaft, Leipzig 1969, p. 301.

GREEN, D. H.; RINGWOOD, A. E.: Mineralogy of peridotitic compositions under upper mantle conditions. Phys. Earth Planet. Inter. **3** (1970) 359–371.

KALININ, V. A.; RODKIN, M. V.: A physical model of deep-focus earthquakes. Isv. Akad. Nauk SSSR, Fisika Zemli 8 (1982) 3–12.

KITTEL, CH.: Einführung in die Festkörperphysik. Akademische Verlagsgesellschaft Geest & Portig K. G., Leipzig 1973, p. 174.

LANDAU, L. D.; LIFSCHITZ, E. M.: Elastizitätstheorie. Akademie-Verlag Berlin, 1965, p. 13.

LEES, A. C.; BUKOWINSKI, M. S. T.; JEANLOZ, R.: Reflection properties of phase transition and compositional Change Models of the 670 km Discontinuity. J. Geophys. Res. **88** (1983) B10, 8145–8159.

LELIWA-KOPYSTINSKI, J.; TEISSEYRE, R.: Constitution of the Earth's Interior. PWN, Warszawa and Elsevier, Amsterdam, Oxford, New York 1984, pp. 248–288.

PARMENTIER, E. M.: A possible mantle instability due to superplastic deformation associated with phase transitions. Geophys. Res. Lett. **8** (1981) 2, 143–146.

POIRIER, J. P.: Plasticity a haute temperature des solides cristallins. Editions Eyrolles, Paris 1976, Chap. 12.

POIRIER, J. P.: On the kinetics of olivine-spinel transition. Phys. Earth Planet. Inter. **26** (1981) 179–187.

POIRIER, J. P.: On transformation plasticity. J. Geophys. Res. **87** (1982) B 8, 6791–6797.

RINGWOOD, A. E.: Composition and Petrology of the Earth's Mantle. Mc Graw-Hill, New York, N.Y., 1975.

STILLER, H.; FRANCK, S.: On the theory of phase transitions in the Earth's mantle. Proc. of the 17th Assembly of the ESC, Publishing House of the Hungarian Academy of Sciences, Budapest 1980, pp. 369–373.

STILLER, H.; FRANCK, S.; MÖHLMANN, D.: Geodynamics and State of the Earth's Interior J. Geodynamics **1** (1984) 79–100.

STOCKER, R. L.; ASHBY, M. F.: On the rheology of the upper mantle. Rev. Geophys. Space Phys. **1** (1973) 391–426.

SUNG, C. M.; BURNS, R. G.: Crystal structural features of the olivine-spinel transition. Phys. Chem. Miner. **2** (1978) 177–197.

TRUBITSYN, C. P.: Phase transitions, isothermal compressibility and thermal expansion of the Earth. Isv. Akad. Nauk SSSR, Fizika Zemli 8 (1979) 21–27.

YODER, H. S., JR.: Generation of Basaltic Magma. National Academy of Sciences, Washington, D.C., 1976.

ZHARKOV, V. N.; TRUBITSYN, V. P.: Physics of Planetary Interiors. (Transl. by W. B. HUBBARD) Pachart, Tucson, AZ, 1978.

The effect of solid phase transformations on the deformation-strength properties of material and their part in seismotectonic processes

V. A. KALININ, I. S. TOMASHEVSKAYA, M. V. RODKIN

The results of geophysical research of high pressure effect indicate that the larger part of rock-forming minerals experience phase transformations at pressures and temperatures existent in the Earth's interior [1]. It was also shown experimentally that during phase transitions the physical properties of solids have certain peculiar features which are absent in stable phases [2]. It was established, for example, that in the process of solid phase transformations the deformability of material sharply increases, as also does the diffusion rate and heat expansion coefficient, whereas the propagation velocities of elastic waves and the strength yield are visibly lower [3, 4].

Since the rocks composing the Earth are mostly polymineral formations, it is important to know when studying geotectonic processes, how the properties of rocks change during the phase transformation in one of the minerals. With this purpose in view, the first part of the paper analyses the deformation-strength properties of a number of rocks with different quartz content during α-β transition in quartz. Quartz is one of the most common rock-forming minerals in which the α-β transition occurs under atmospheric pressure and a temperature of 573 °C. In the second part of the paper, the anomalous properties of rocks during phase transformations are used for physical interpretation of certain seismotectonic processes.

The deformation-strength properties of rocks were experimentally studied by the uniaxial compression of samples having the shape of the rectangular parallelepiped with the sides $2 \cdot 2 \cdot 5$ cm. Compression was applied along the longest side of the parallelepiped. The samples were prepared from quartzite (the monomineral quartz rock), granite (30% of quartz), altered gabbro (3% of quartz), and peridotite (quartzless rock).

There were three sets of experiments.

1. The isothermal stress-deformation curves were plotted for different temperatures. thus allowing to obtain the dependence of the rock strength yield on temperature.
2. The dependence of the sample's length on temperature was studied under constant uniaxial loading.
3. The mechanical state of the heating inclusion into the rock mass was simulated in the third set of experiments.

The first set of experiments established that the quartz containing rocks are weakest in the temperature range of the α-β transition in quartz. At these temperatures the strength yield is twice as low as at room temperature. Moreover, the sample failure in the temperature range of solid phase transformation produces a large amount of finely fragmented material practically absent in the destroyed samples beyond the temperature range of the solid phase transformation. As an example, Fig. 1 gives the dependence of the strength yield of altered gabbro (3% of quartz) on temperature and describes schematically the destruction process of samples at different temperatures.

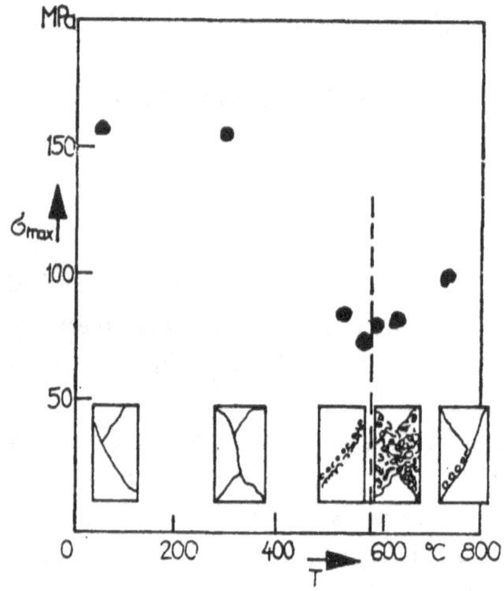

Fig. 1. Dependence of strength yield of altered gabbro on temperature and the nature of failure of samples.

The results of the second set of experiments are shown in Fig. 2 by the temperature dependence of sample-lengthening made of different materials under constant uniaxial loading of 72—73 MPa. In peridotite (quartzless rock) the lengthening of the sample due to thermal expansion occurs in the entire studied temperature range reaching 720 °C. Against the background of sample lengthening due to thermal expansion at temperatures of 581—587 °C, quartzite shows a short-term reduction of sample length ascribed to the α-β transition in quartz. Though this sample was heated to 700 °C and experienced loading, there was no evidence of failure.

Against the background of the thermal expansion of the altered gabbro samples, the anomalous change of length occurred in a wider temperature range. This probably has two reasons; firstly, the α-β transition in individual quartz grains may take place at different temperatures because of irregular stress throughout the sample, and, secondly, the phase transformation in quartz may possibly encourage cracking in the sample followed by destruction. At lower stress levels, the samples of altered gabbro were not destroyed and the changes in the length of the sample had a more gentle sloping minimum than quartzite. In granite the phase transformation shifted towards higher temperatures apparently owing to the change in the stress state of quartz grains, and the sample deteriorated into small pieces.

In the third set of experiments we tried to create conditions under which a certain volume of inclusion could exist inside the rock mass. This inclusion contained quartz, whereas the enclosing rocks were quartzless. The scheme of the experiment was as

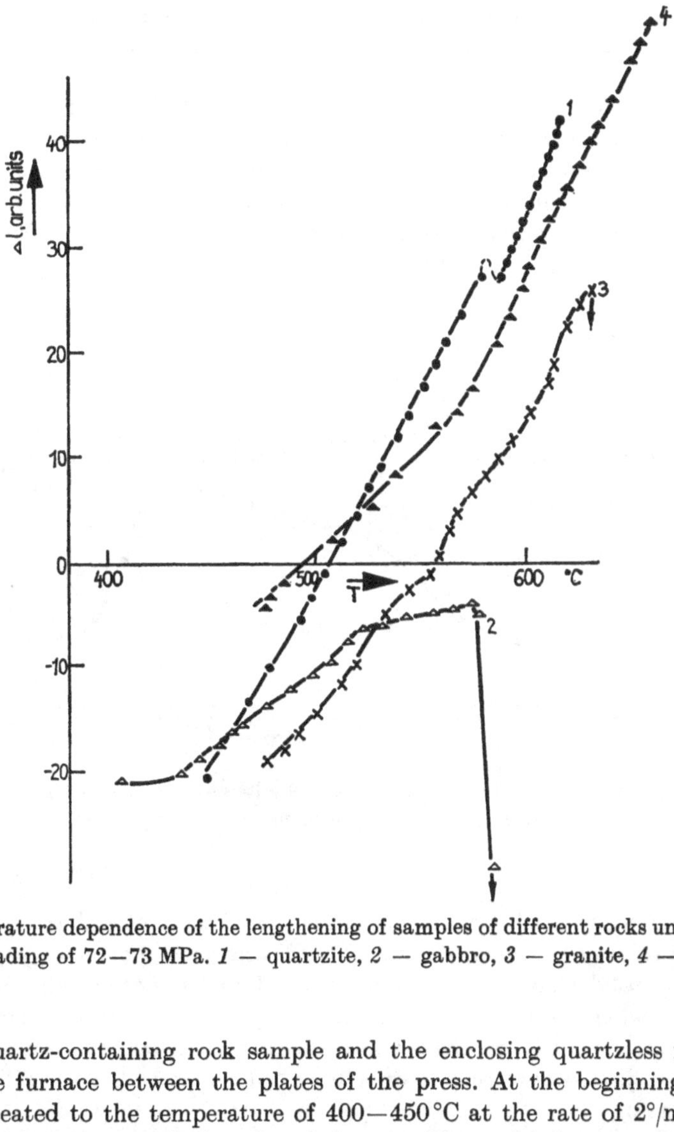

Fig. 2. Temperature dependence of the lengthening of samples of different rocks under continuous uniaxial loading of 72—73 MPa. *1* — quartzite, *2* — gabbro, *3* — granite, *4* — peridotite.

follows: a quartz-containing rock sample and the enclosing quartzless framing were placed in the furnace between the plates of the press. At the beginning, the sample was slowly heated to the temperature of 400—450 °C at the rate of 2°/min, and then a compressing force was applied to the whole assemblage; this force was approximately equal to one half of the force, which destroy the quartz inclusion, and the position of the plates of the press were fixed. With the rise of temperature we observed the change in the force due to processes occurring in the sample-inclusion and in the enclosing framing and the changes in the length of sample and the adjoining parts of the enclosing framing. Fig. 3 shows the resulting pattern. In the region of temperatures from 550 to 580 °C, the anomalous changes in force and length are recorded, which can be attributed to anomalously great volumetric expansion of quartz when approaching phase transformation and to the increase of the deformability of rock on the whole during the phase transition.

These results imply that the processes of phase transformation, and possibly of other solid phase transformations as well, such as dehydration, decomposition, etc., impart certain peculiarities to the mechanism of preparation of failure and to destruction itself.

The experiments described show that not only minerals, but rocks likewise manifest anomalous deformation-strength properties during solid phase transformations. Under these conditions, the rocks lose their strength and become more plastic. This means that from the mechanical viewpoint the phase boundaries in the Earth are weakened zones [5]. For example, relative displacements can easily occur along these zones and earthquake foci are often located there.

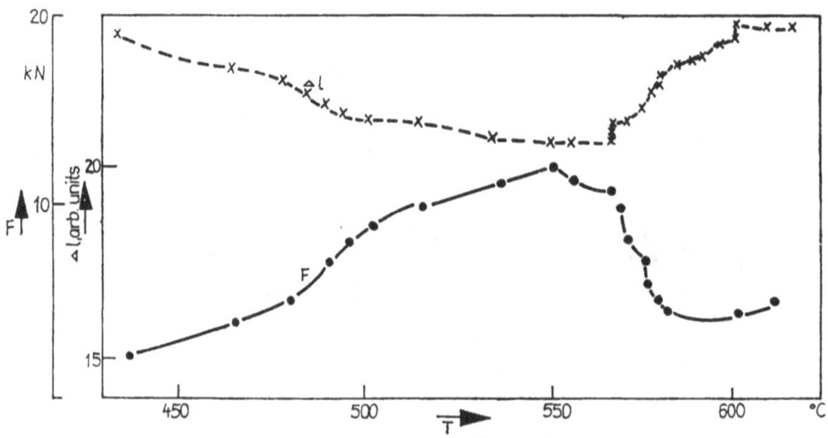

Fig. 3. Results of "model" experiment: enclosing rock-altered gabbro. Solid line-change of forces, dashed line-change of length under rise of temperature.

Bearing in mind, that the anomalous properties of the material are evident only in the process of solid phase transformations, the layers with anomalous physical properties, caused by solid phase transformations, are found in tectonically active regions, where the changing thermodynamic conditions produce phase and chemical transformations of material, though phase boundaries are observed universally in the Earth.

We shall now discuss how the concepts described correspond to the different seismotectonic processes.

During the last few years, the concept of tectonic layering of the lithosphere has been intensively developed in geology [6]. According to this concept, the lithosphere of tectonically active areas consists of relatively hard quasi-horizontal plates of different thickness. The plates are divided by mechanically weakened and easily deformed layers, on which the plates move against each other. The horizontal displacements along these layers are on the average about 100 km, the largest observed ones reaching up to 500—600 km. In several regions, the weakened layers are confined to the fronts of metamorphism.

This concept is based on geological data. The application of geophysical methods established the presence of quasi-horizontal layers with higher seismicity and lower seismic wave velocities. In a number of cases, the displacement planes in the earthquake foci in the subhorizontal layers with higher seismicity are oriented along these layers.

The averaged values in different regions indicate that the low velocity layers LVZ are mainly located at the depths of 10—13 km, 21—23 km, and 38—44 km. The largest number of layers with higher seismicity are located at nearly depths (9—13 km, 22 to 27 km, and 40—45 km). Let us note, that the significant zones of transformation of material in the lithosphere are usually related to the same depths: green-shale metamorphism (9—14 km), granitisation (20—25 km), and the average depth of the Moho boundary (35—40 km). This coincidence is probably not accidental.

In view of this, the essential part of the concept of tectonic layering of the lithosphere is the notion about mechanically weakened layers confined to the fronts of metamorphism with higher seismicity and lower velocities of elastic waves propagation. The genesis of these layers is as yet unknown. The most popular opinion suggests melting in these layers. This process can apparently explain many of the enumerated empirical characteristics of weakened layers. However in the light of the latest data on the thermal regime of the lithosphere an extensive melting at the given depths seems hardly possible.

Let us discuss the alternative view on the genesis of mechanically weakened layers in tectonically active regions, which is based on the anomalous properties of the material observed during the microstructural changes of the material.

The anomalous physical properties of material provide the natural explanation of all empirical peculiarities of mechanically weakened layers, which form the basis of conception of tectonically layered lithosphere.

Indeed we have just described the reason of confinement of these layers to tectonically active regions. The reduction of elastic waves velocities during the solid phase transformation correlates with the low velocity layers observed in nature in weakened layers. A lower strength yield during transition indicates that the failure of material more readily occurs there, which means that earthquakes should take place in these layers more frequently and, therefore, they have higher seismicity. Finally, in the presence of a corresponding system of mechanical stresses a higher deformability of the material in these layers causes the harder lithospheric plates to slip along one another.

We shall now briefly analyse the nature of deep-foci earthquakes. This kind of earthquakes occurs in the Benioff zones, at the depth of more than 200 km. There are difficulties in the physical interpretation of processes in the sources of such earthquakes. That is why Bridgman suggested, in as far back as 1945, that the earthquakes are related to phase transformations. The physical nature of this relation, however, is still obscure, thus giving birth to a variety of models of interpretation of the connection of earthquakes with phase transformations. We shall not dwell on the detailed analysis of these models, but remark that they ascribe the major cause of earthquakes to the density jump during phase transformation. In this case, the radiation from the source should be similar to the radiation from an explosion, which can hardly be correlated with displacements in the focus.

In 1981 two of the authors of this paper published a new model of deep-foci earthquakes based on the anomalous deformability of materials manifested in the process of phase transformation [5]. The principal idea of the model was as follows.

In accordance with the basic concepts of plate tectonic, the subducting plate is

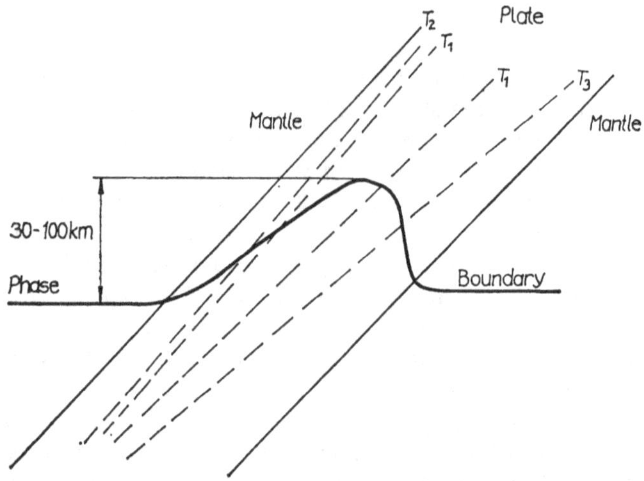

Fig. 4. The phase boundary position inside the subducting slabe.

colder than the surrounding mantle. As shown by recently obtained data, since the tilt of the phase curve on P, T plane of the major phase transformations in the upper mantle is positive, the phase boundary horizontal in the mantle within the plate deviates upward and has the shape of a swell, see Fig. 4. The material of the plate during its subsidence crosses the phase boundary and from the low pressure phase turns into the high pressure phase. In the process of transformation, the material becomes plastie and the mechanical stresses in the plate may cause sudden displacement along the surface of the phase interface.

Fig. 5 shows the vertical profiles of several Benioff zones. It should be noted that at certain depths the lithospheric plates sharply change the angle of their tilt with respect to the horizon. In these places the deformations may reach a few tens of per-

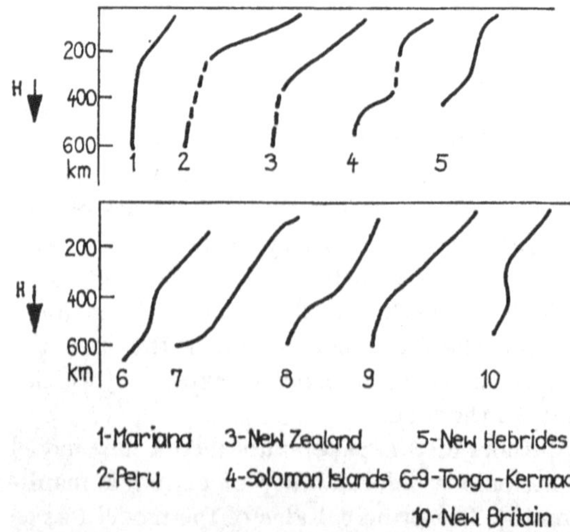

Fig. 5. Morphology of several Benioff zones from the data in the literature.

cent. In the location of the bend the material of the plate ought to be highly plastic to reach such large deformations. The depths, where the lithospheric plates sharply bend, are in good correlation with the depths of the major phase transformations and with the maximums of deep-foci seismicity.

Therefore, the anomalously high plasticity of the material of the plate is apparently caused by high deformation capacity of this material during phase transformation. Fig. 6 shows schematically the process of bending of the subducting plate; dashed arrows indicate the directions of slipping of the upper and lower parts of the plate.

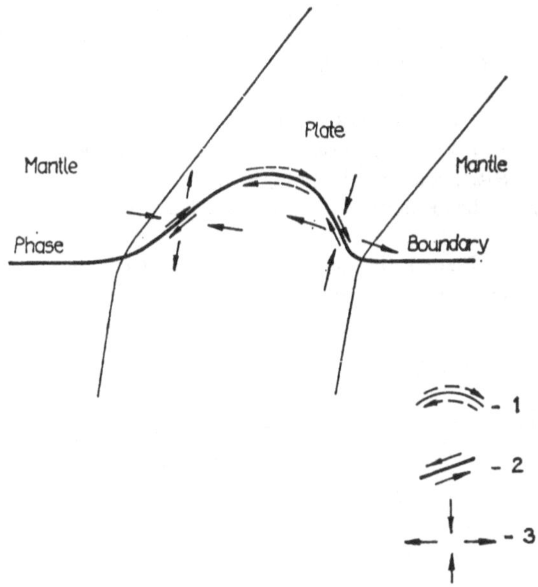

Fig. 6. The scheme of bending of the subducting plate: dashed arrows show the direction of slipping of parts of the plate; half-arrows show the direction of displacement in two earthquake foci; arrows show the direction of major compression and tension axes.

Consequently, the deep-foci earthquakes and the sudden bends of lithospheric plates are caused by displacements along one and the same surface of the phase interface. The half-arrows in Fig. 6 denote the directions of displacements in two earthquake sources. The arrows also show the directions of the major compression and tension axes of seismic focal mechanisms of these earthquakes.

The model of seismotectonic processes in the subsiding lithospheric plates discussed above was used for the analysis of deep-foci earthquakes in the Kuril-Kamchatka seismic zone. The results showed that almost all known focal mechanisms of deep-foci earthquakes, which occurred in this area, satisfy the theoretical model described. In particular, these results challenge the statement made by Molnar and Isaks that at certain depths the lithospheric plates are compressed in the direction of subsidence, whereas in other directions they are tensioned [7]. As shown in Fig. 6, only the inner edge of the plate is compressed while the outer edge is tensioned.

In conclusion we may add, that in this paper we endeavoured to demonstrate not only the new actual physical models of seismotectonic processes at different depths in the Earth, but the important part of solid phase transformations in the interpretation of macroprocesses active in the Earth, and the necessity of further detailed research of various properties of material during such transformations.

References

[1] RINGWOOD, A. E.: Phase transformations in descending plate and implications for mantle dynamics. Tectonophysics (1976) 1–2, 129.

[2] POIRIER, J. P.: On transformational plasticity. J. Geophys. Res. 87 (1982) 38, 6791.

[3] VOLAROVICH, M. P.; BAYUK, E. I.; EFIMOVA, G. A.: Elastic properties of minerals under high pressures (in Russian). Nauka (1975) 123.

[4] KALININ, V. A.; TOMASHEVSKAYA, I. S.: Plasticity of minerals during phase transformations (in Russian). Dokl. AN SSSR 268 (1982) 1, 59.

[5] KALININ, V. A.; RODKIN, M. V.: The physical model of sources of deep-foci earthquakes (in Russian). Izv. AN SSSR, Fizika Zemli (1982) 8, 3.

[6] Tectonic layering of the lithosphere in recent mobile belts (in Russian). Coll. of papers, Nauka (1982).

[7] ISAKS, B.; MOLNAR, P.: Distribution of stress in descending lithosphere from a global survey of focal-mechanism solutions of mantle earthquakes. Rev. Geophys. Space Phys. 9 (1971) 1, 103.

The quantitative deep earthquakes model

V. A. Kalinin, M. V. Rodkin

The qualitative deep earthquakes model based on the significant increasing of the plasticity during the solid phase transformations was proposed by authors earlier [1]. As was supposed the deep earthquakes are taking place during the fast avalanch like phase transition of the metastable phase. The metastable phase can be formed in the subducting lithospheric plates due to the sufficiently low temperatures of these plates. The abrupt plasticity increase caused from the solid state transformation leads to the origin of the fluid like layer along the phase boundary in the subducting slab. The direction of the sliding along this layer is determined by the acting tectonic stresses. The fast sliding can be registered as the deep earthquake. According to the proposed model the deep earthquake mechanism should include two components: volume, caused by the phase transformation with the bulk change and shear one, arising due to sliding along the mechanically weakend phase boundary.

The supposed model agrees well with the geophysical data. The shear mechanisms with small volume part are typical for deep earthquakes [2, 3]. The volume strains are dominant only in very rare earthquake mechanisms. Besides, according to [4], the deep earthquakes cease if the temperatures of the subducting plates exceed 800 to 900 °C. As was discussed by A. E. Ringwood [5] the phase transformations in silicates run sufficiently quickly when temperatures exceed 800–900 °C. So the metastable phase has no origin and according to our model the deep earthquakes cannot take place. The more detailed comparison of the proposed model with geophysical data is described in [6–8]. The accordance is sufficiently good.

The described model is qualitative. Now we suppose the quantitative description of this model.

According to the experimental and theoretical research [9, 10], the deformation velocity $\dot{\varepsilon}$ of the specimen undergoing the phase transition with the volume changing $(\Delta V/V)$ can be described as

$$\dot{\varepsilon} = R\left(\frac{\Delta V}{V}\right)\left(\frac{\sigma}{J}\right)\dot{c}, \tag{1}$$

where R — coefficient; σ — deviatoric stress; J — yield point; $\dot{c}$ — the transformation velocity (c — one of the phase concentration).

The R value is near 1 when $\sigma \ll J$ and will quickly increase with σ increasing when $\sigma \gtrsim J$ [11]. This raising of $\dot{\varepsilon}$ can be taken into account by effective increasing of R

The effective value of R when $\sigma \gtrsim J$ is near 10. One can see from (1) that high $\dot{\varepsilon}$ meanings can be reached only when $\dot{c}$ is very high too. Such values of $\dot{c}$ can result during the avalanch like phase transition stage. So the avalanch like phase transition stage is necessary for the deep earthquakes origin. The conditions essential for the formation of this stage will be discussed below.

The kinetic equation of the phase transition in solids can be written as [12]

$$\dot{c} = R_0 \left(\frac{kT}{h}\right) \exp\left(-\frac{\Delta E}{kT}\right) \left\{\exp\left(\frac{\mu_1 - \mu_2}{kT}\right) - 1\right\} (1 - c), \tag{2}$$

where R_0 — coefficient; k — Boltzmann constant; T — temperature; h — Plank constant; ΔE — activation energy; μ_1, μ_2 — Gibbs thermodynamic potentials of the phases.

During the fast stage of the phase transformation the process is adiabatic. In this case we have

$$T = T_0 + \frac{Q_c}{C_p} + C\delta T_p, \tag{3}$$

where T_0 — initial temperature; Q — latent heat of the phase transition; C_p — specific heat; δT_p — the temperature increasing due to the energy dissipation during the phase transition.

The avalanch like phase transition stage is more probable at the beginning of the phase transformation. Thus

$$c \ll 1. \tag{4}$$

From (3), (4) one have

$$\exp\left(-\frac{\Delta E}{kT}\right) = \exp\left(-\frac{\Delta E}{kT_0}\right) \left\{1 + \frac{Q_c \Delta E}{kC_p T_0^2} + \frac{c\,\Delta E \delta T_p}{kT_0^2}\right.. \tag{5}$$

Since

$$|\mu_1 - \mu_2| \gtrsim kT,$$

that

$$\exp\left(\frac{\mu_1 - \mu_2}{kT}\right) \simeq 1 + \left(\frac{\mu_1 - \mu_2}{kT}\right) + \frac{1}{2}\left(\frac{\mu_1 - \mu_2}{kT}\right)^2. \tag{6}$$

The difference of the Gibbs potentials changing due to the phase transition can be described as [13]

$$\mu_1 - \mu_2 = (\mu_1 - \mu_2)_0 - \frac{Q\delta T}{T} - \Delta V\,\delta P, \tag{7}$$

where $(\mu_1 - \mu_2)_0$ — Gibbs potentials difference at $t = 0$; $\delta T, \delta P$ — T, P variations due to the phase transition. These variations lead to the decreasing of the initial $(\mu_1 - \mu_2)_0$ value.

The δT variations have been described by [3]. The δP values are difficult to define exactly because δP may change in the wide limits

$$0 < \delta P < \frac{\Delta V}{V} Kc, \tag{8}$$

where K is a bulk module.

The first case — minimum value of δP — corresponds to the ideal plastic local stress relaxation. The maximum one — to the elastic solid. We will examine both cases.

For the first case from (3), (4), (6)—(8) one has

$$\exp\left(\frac{\mu_1-\mu_2}{kT}\right)-1\simeq\frac{(\mu_1-\mu_2)_0}{kT_0}-\frac{Q^2c}{kC_pT_0^2}-\frac{Q_c\delta T_p}{kT_0^2}+\frac{1}{2}\left\{\frac{(\mu_1-\mu_2)_0}{kT}\right\}^2 - \frac{(\mu_1-\mu_2)_0}{kT_0}\left\{\frac{Q^2c}{C_pT_0}+\frac{Q_c\delta T_p}{T_0}\right\}. \tag{9}$$

After the substitution (5) and (9) in (2) one has an equation which may be rewritten as

$$\dot{c}=a+bc, \tag{10}$$

where a and b are coefficients depending on the phase transition properties and thermdynamic conditions.

From (10) it is easy to see that the avalanch like phase transition stage takes place when $b>0$.

In this case one can obtain the condition of the avalanch like stage generating

$$(\mu_1-\mu_2)_0\left\{\frac{Q\Delta E}{kC_pT_0^2}+\Delta\frac{E\delta T_p}{kT_0^2}-1\right\}\left(1+\frac{(\mu_1-\mu_2)_0}{2kT_0}\right) - \left(\frac{Q^2}{C_pT_0}+\frac{Q\delta T_p}{T_0}\right)\left(1+\frac{(\mu_1-\mu_2)_0}{kT_0}\right)>0. \tag{12}$$

If considering only the linear part in (6) one receives the more simple condition

$$(\mu_1-\mu_2)_0\left\{\frac{Q\Delta E}{kC_pT_0^2}+\frac{\Delta E\delta T_p}{kT_0^2}-1\right\}-\frac{Q^2}{C_pT_0}-\frac{Q\delta T_p}{T_0}>0. \tag{12'}$$

For the second case (elastic solid) we have

$$(\mu_1-\mu_2)_0\left\{\frac{Q\Delta E}{kC_pT_0^2}+\frac{\Delta E\delta T_p}{kT_0^2}-1\right\}\left(1+\frac{(\mu_1-\mu_2)}{2kT_0}\right) - \left(\frac{\Delta V^2k}{V}+\frac{Q^2}{C_pT_0}+\frac{Q\delta T_p}{T_0}\right)\left(1+\frac{(\mu_1-\mu_2)_0}{kT}\right)>0 \tag{13}$$

and accordingly for more simple linear version of (6)

$$(\mu_1-\mu_2)_0\left\{\frac{Q\Delta E}{kC_pT_0^2}+\frac{\Delta E\delta T_p}{kT_0^2}-1\right\}-\left(\frac{\Delta V^2k}{V}+\frac{Q^2}{C_pT_0}+\frac{Q\delta T_p}{T_0}\right)>0. \tag{13'}$$

The received equations are bulky enough. Therefore, we shall discuss more simple linear cases (12′) and (13′) only. From these conventions it is easy to see that the avalanch like phase transition stage is caused by the phase transition speed up due to the kinetic effect of the temperature increasing. At some conditions this speed up of the phase transition can prevail its going-down due to $(\mu_1-\mu_2)$ reducing through time.

According to (12′) and (13′) the avalanch like phase transition stage is possible during the transitions with ΔE sufficiently large when the initial temperature T_0 is small enough. Besides, the avalanch like stage generating can be promoted by the $(\mu_1 - \mu_2)_0$ increase.

The received conclusions are confirmed by the experimental data. The quick jumps of the pase transitions have been registered at moderate and lower temperatures during the phase transitions with the large values of ΔE [12, 14].

However, as was discussed earlier, generating of the avalanch like phase transition stage is necessary but insufficient for the deep earthquake origin. The avalanch like stage may cease too soon and does not lead to the high values of $\dot{c}$. The determined values of $\dot{\varepsilon}$ will be too small to be registered as deep earthquake. Therefore, we have to discuss the probable velocities of the sliding during the mechanically weakened phase boundary.

From (1) we conclude that the solid undergoing phase transition is rheological similar to the viscous fluid with the effective viscotity

$$\nu = \left\{R \frac{\Delta v}{v} \cdot \frac{\dot{c}}{I}\right\}^{-1}. \tag{14}$$

Thus the phase transformations layers in solids may be described as viscous fluid layers. The blocks separated by the mechanically weakened phase boundary may slide along this fluid like layer with the velocity

$$V = \frac{\sigma \delta h}{\nu}, \tag{15}$$

where δh is a layer thickness; σ is the tectonic stress acting along the phase boundary.

From (14) and (15) one has

$$V = R \frac{\Delta v}{v} \frac{\sigma}{I} \dot{c} \delta h. \tag{16}$$

We shall discuss the quick avalanch like stage of the phase transition. Let this stage run during the time interval τ. So for the rough examination we suppose

$$\dot{c} \simeq 1/\tau \tag{17}$$

and

$$\delta h \simeq V_{ph} \tau, \tag{18}$$

where V_{ph} is the phase boundary displacement velocity. From (16), (17), (18) one has

$$V \simeq R \left(\frac{\Delta v}{v}\right) \left(\frac{\sigma}{I}\right) V_{ph}. \tag{19}$$

The existence of the wide-spread plastic deformations in the earth crust allows us to suppose $\sigma \gtrsim I$. For such σ the effective R values are near 10.

The volume changing $(\Delta V/V)$ for the main crust and upper mantle phase transitions is near 10%. Thus we obtain from (19)

$$V \simeq V_{ph}. \tag{20}$$

Hence using the V_{ph} values registered for the solid phase transitions with high $(\Delta V/V)$ meanings [12] one can receive $V \simeq 1 + 10^2$ cm/s. The received estimation is in the good accordance with the geophysical data [15].

Using this model one can also compare the volume and shear components of the deep earthquake focal mechanism. The material displacement velocity arising due to the volume changing at the phase transition can be estimated as $V_v = (\Delta V/V) \cdot V_{ph}$. Thus the relation of the volume deformation velocity V_v to the shear one V_{sh} can be written as

$$\frac{V_v}{V_{sh}} \simeq \frac{\left(\frac{\Delta V}{V}\right) V_{ph}}{R\left(\frac{\Delta V}{V}\right)\left(\frac{\sigma}{I}\right) V_{ph}} = \frac{I}{\sigma R}. \tag{21}$$

Using the earlier discussed estimations $\sigma \gtrsim I$, $R \simeq 10$ one gets from (21) $V_v/V_{sh} \simeq 10\%$. Thus the shear component of the deep earthquake focal mechanism is dominant. However, the shear component is quickly decreased when σ decreases. Therefore in the regions where the tectonic stresses σ are sufficiently small the deep earthquakes with the volume type focal mechanism can occur. The deep earthquakes with dominant volume component of the focal mechanism have been registered in some cases.

The received numerical estimations support the qualitative model of the deep earthquakes based on the abrupt increasing of the plasticity in solids during the phase transitions as has been proposed by authors earlier.

References

[1] Kalinin, V. A.; Rodkin, M. V.: Izv. AN SSSR, Fizika Zemli (1982) 8, 3.

[2] Randall, M. J.; Knopoff, L.: J. Geophys. Res. **75** (1970) 26, 4965.

[3] Wyss, M.; Molnar, P.: Phys. Earth. Pl. Int. **6** (1972) 4, 279.

[4] Molnar, P.; Freedman, D.; Shih, J. S. F.: Geoph. J. Roy astr. Soc., **56** (1975) 1, 41.

[5] Ringwood, A. E.: Composition and petrology of the Earth's mantle, McGraw-Hill, Inc., 1975.

[6] Kalinin, V. A.; Rodkin, M. V.: Doklady AN SSSR **284** (1985) 1, 87.

[7] Rodkin, M. V.: In: Aktualnye problemy geophysici, Moskow, Nauka (1985) 115.

[8] Kalinin, V. A.; Rodkin, M. V.: Izv. AN SSSR, Fizika Zemli (1986) 1, 3.

[9] Poirier, J. P.: J. Geophys. Res. **87** (1982) 39, 6791.

[10] Andronov, N. N.; Kuzmin, S. L.; Likhachev, V. A.; Patrikeev, Yu. J.; Korolev, M. N.: Problemy prochnosti (1983) 5, 96.

[11] Greenwood, G. W.; Jonhson, R. H.: Proc. Roy. Soc. A. **283** (1965) 1394, 403.

[12] Christian, J. W.: Transformations in Metals and Alloys, Part 1., Pergamon Press, 1975.

[13] Sirota, N. N.: In: Kristallizasia i phazovye prevrashchenia. Nauka i tekhnika, Minsk, 1971, 333 p.

[14] Livshis, L. D.; Riabinin, Yu. N.: In: Phizicheskia osnovania poiskov metodov prognoza zemletrasenii. Moskow, Nauka (1970) 28.

[15] Moskvina, A. G.: Izv. AN SSSR, Fizika Zemli (1984) 7, 38.

The effect of the α-β-transition on the thermal diffusivity of quartz-containing rocks

U. Seipold

For pure anorganic compounds the existence of phase transitions has an important influence on the thermal conductivity curve. For example in Fig. 1 the behaviour of potassium bromide at the pressure induced B1—B2 structure change is shown. The thermal conductivity is lowered by about 40 per cent in the region of the crystal transformation. We intended to investigate the influence on the phase transition effect, if the mineral with the structure change is placed in the framework of a natural rock sample. For this investigation we have chosen the α-β-transition of quartz. The trigonal structure of the α-quartz is being transformed into the hexagonal structure of the β-modification at 575 °C. The transition temperature has a strong pressure dependence. The pressure coefficient has the value of 260 K/GPa [Yamamoto, 1974].

Several authors have investigated the influence of the structure change on the petrophysical properties [Coe, 1969; Ghiorso, 1979; Moore, 1979 and Kern, 1982]. In the measurements of the thermal conductivity or diffusivity for quartz-containing rocks the effect of the transition could not be observed. The reason for this was that the distance of the measuring points was too great. Therefore, we measured with a higher

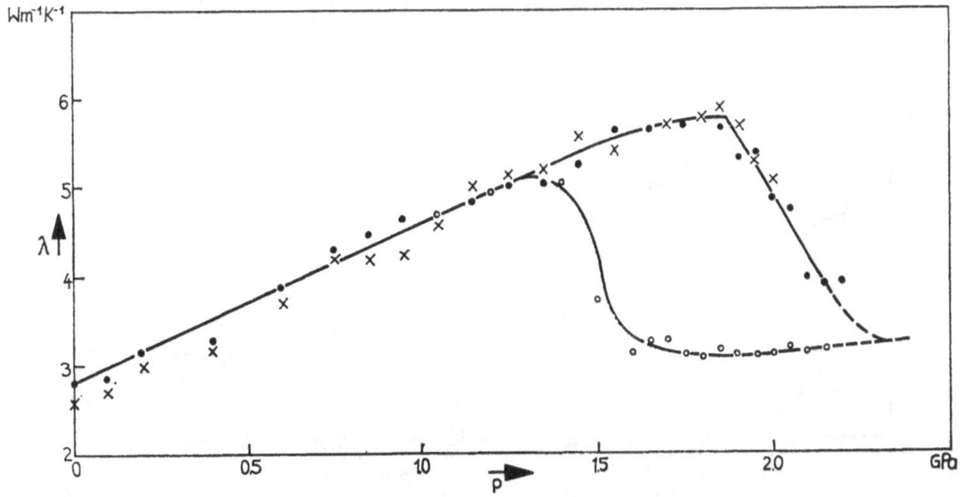

Fig. 1. Thermal conductivity at the phase transition for KBr.

number of measuring points in the temperature interval of interest. The investigation was performed by using the modified Angstroem method as described by SEIPOLD [1980].

The high pressure was generated by means of a cubic press. For the measurements fine grained granulites were used with high quartz content.

For the mineral content see Table 1.

Table 1. Mineral composition in percent

	quartz	plagio-clase	potash feldsp.	garnet	biotite	kyanite
sample 8, 9	46.8	22.6	16.1	2.5	9.8	1.9
sample 10	38	9	47	4	1	—

Under these conditions the influence of the transition on the thermal diffusivity of the investigated granulites was clearly detectable. The best way to make the effect evident was to represent the results on a logarithmic scale (see Figs. 2, 3). The thermal diffusivity went through a minimum and increases again in the region of the high pressure phase. This behaviour corresponds to the properties of the elastic velocities as observed by FIELITZ [1976] and KERN [1979]. Therefore, the minimum can be explained by a decrease of the structure change. But the same effect can also be obtained by lowering the mean free path length caused by crystal imperfections generated by the structure transition. We believe that both reasons work together in the same direction.

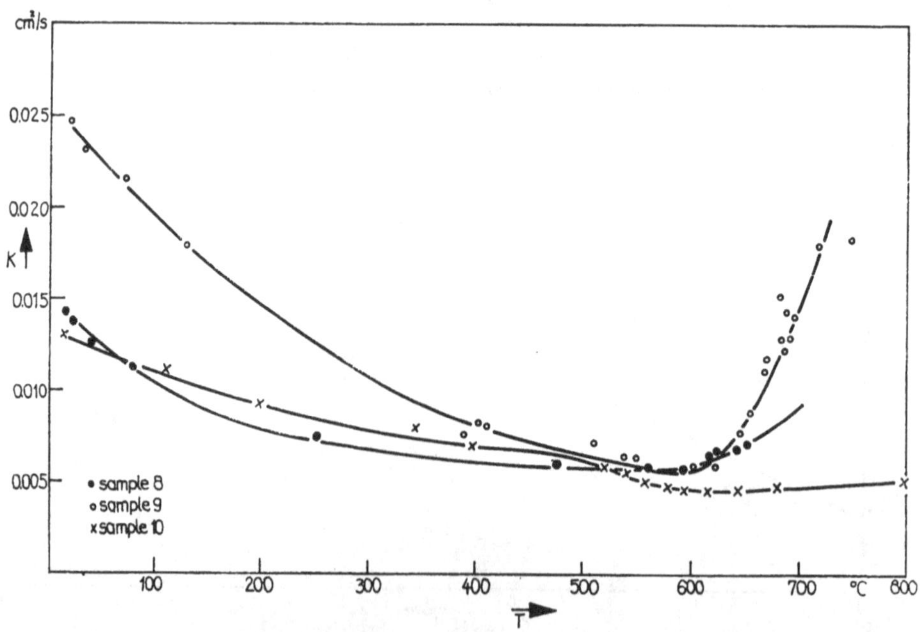

Fig. 2. Thermal diffusivity at the α-β-transition.

The increased thermal diffusivity of the high pressure phase is in coincidence with the Debye theory. The measurements were carried out at a pressure of 100 MPa under quasihydrostatic conditions with varying temperature. For this pressure the transition should occur at 601 °C but actually it is shifted to samewhat higher temperatures. The reason for this could be local stress differences caused by the anisotropic expansion of the mineral grains when the rock sample is heated. VAN DER MOLEN [1981] stated that for this reason the pressure coefficient in rocks can be increased from 26 K/100 MPa up to 60 K/100 MPa. The displacement of the transition to higher temperatures was also observed by FIELITZ [1976] investigating the velocities of elastic waves. The phase transition appears as an sensitive indication for the actual stresses in the rock sample.

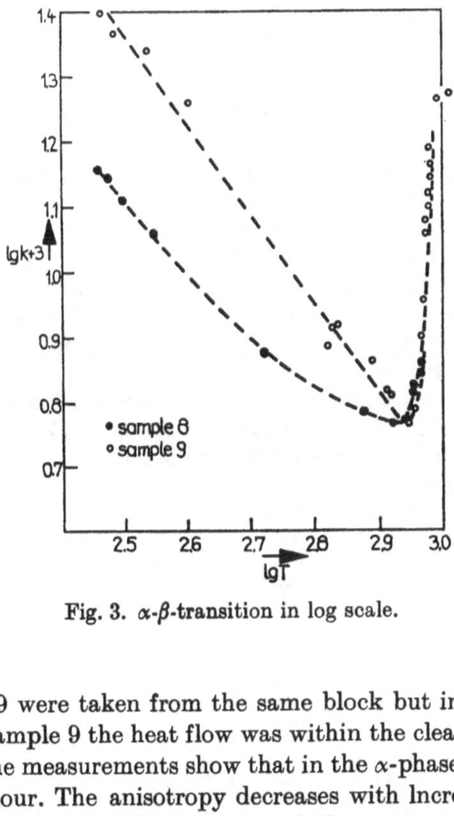

Fig. 3. α-β-transition in log scale.

The samples 8 and 9 were taken from the same block but in perpendicularly orientated directions. For sample 9 the heat flow was within the clearly visible metamorphic plane. The results of the measurements show that in the α-phase the thermal diffusivity has anisotropic behaviour. The anisotropy decreases with lncreasing temperature and vanishes at the transition point. There are some different reasons for the anisotropy in quartz-containing rocks. At first, the quartz has a higher conductivity than the other parts of the rock. Therefore in the direction of the longer axis of the quartz grains we find a higher conductivity.

Secondly, each individual quartz grain has a strong anisotropy. Mostly the longer axes of the grains are in coincidence with the c-axes of the crystal. In this direction the conductivity of the crystal is higher than in the direction of the a-axis by a factor 1.7 [HORAI, 1969]. At third, in a direction perpendicularly to the long grain axis the number of grain boundaries is higher and the conductivity is lower therefore. All three reasons for the anisotropy work in the same sense for the special case of quartz.

For other minerals we would find an other situation. From the disappearance of the anisotropy at the crystal transformation point we can follow that the crystal anisotropy plays the dominationg role.

Summarising one can say that due to the influence of the rock framework the effect of the α-β-transition of the quartz grains is shifted and distributed over a greater temperature range. In this way it does not change considerably the thermal diffusitity versus temperature curve.

References

Coe, R. S.; Paterson, M. S.: The α-β-inversion in quartz. J. Geophys. Res. **74** (1969), 4921–4948.

Ghiorso, M. S.; Carmichael, J. S. E.; Moret, L. K.: Inverted high temperature quartz. Contr. Miner. Petrol. **68** (1979), 307–323.

Horai, K.; Simmons, G.: Thermal conductivity of rockforming minerals. Earth Planet. Sci. Letters **6** (1969), 359–368.

Kern, H.: Elastic wave velocity in crustal and mantle rocks at high pressure and temperature. Phys. Earth Planet. Inter. **29** (1982), 12–23.

Kern, H.: Effect of high-low quartz transition on compressional and shear wave velocities in rocks under high pressure. Phys. Chem. Miner. **4** (1979), 161–167.

Moore, G. S. M.; Rose, H. E.: Thermal effects of contamination, absorbed water and annealing on the DTA of powdered quartz. J. therm. Anal. **15** (1979), 37–45.

Seipold, U.; Gutzeit, W.: Measurements of the thermal properties of rocks under extreme conditions. Phys. Earth Planet. Inter. **22** (1980), 272–276.

Van der Molen, J.: The shift of the α-β-transition temperature of quartz associated with the thermal expansion of granite at high pressure. Tectonophysics **73** (1981), 323–342.

Yamamoto, A.: Lattice dynamical theory of structural phase transition in quartz. J. Phys. Soc. Japan **37** (1974), 797–808.

Fielitz, K.: In: Giese, P.; Prodehl, C.; Stein, A.: Explosion Seismology in Central Europe. Springer Berlin, New York 1976.

Investigation of pressure-induced phase transformations by energy-dispersive X-ray diffraction

A. KRAFT, M. ZIEMANN

Introduction

At the Central Institute for Physics of the Earth in Potsdam the field of high pressure physics has been extended. The development of new methods for high pressure research in cluding X-ray diffraction under high pressure is one of the main topics. Especially the diamond anvil high pressure cell has found extensive application in a wide variety of high pressure experiments. Structural studies are being carried out to very high pressure (up to 300 kbar) by using diamond anvil high pressure cells. Powder dif-

Fig. 1. High pressure diamond anvil cell for X-ray diffraction and optical measurements. Anvil faces 0.3—0.4 mm (octagonal), $P_{max} \approx 400$ kbar.

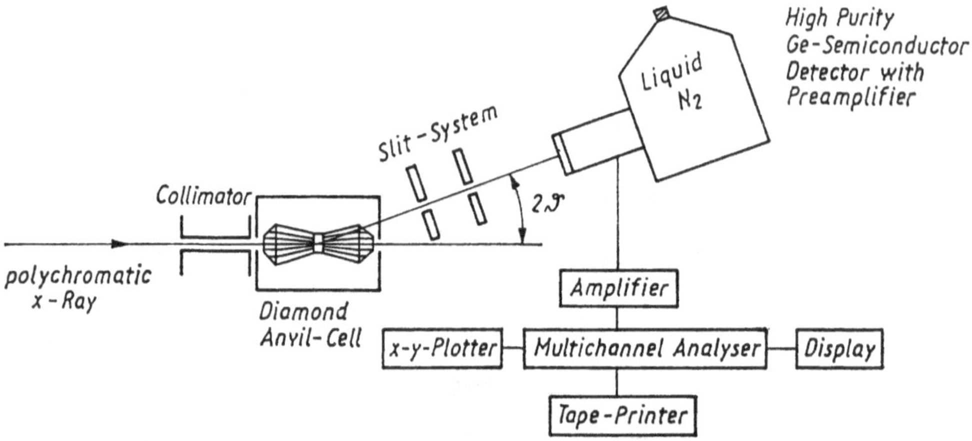

Fig. 2. Block diagram of the experimental set-up for energy-dispersive X-ray diffraction.

fraction patterns have most frequently been measured with monochromatic incident radiation (e.g. filtered MoK_α) and photographic detection. As a result of the small size of the sample and to a lesser extent of the absorption of the beam due to the diamond anvils and sample, such an experiment typically takes one to three days of exposure. Even faster experiments can be carried out by use of energy- dispersive X-ray diffractometry. This method is favourable for crystallographic studies under high temperature and/or high pressure conditions.

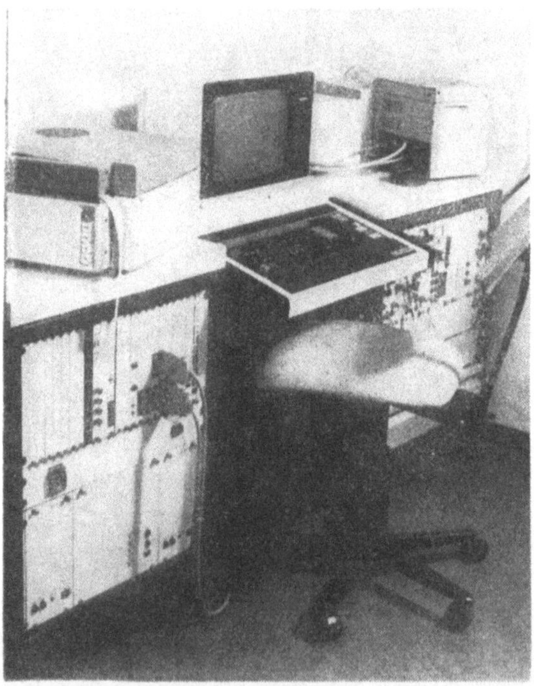

Fig. 3. Multichannel analyser with a microcomputer unit produced in the Centre for the Production of Scientific Instruments of the Academy of Science of the GDR (VKA — 4995).

In energy-dispersive diffraction a beam of polychromatic (white) X-rays passes axially through the first diamond, enters the small sample (0.1—0.2 mm diameter) and the diffracted beam exists via the second diamond.

The wavelength (photon energy) spectrum of the scattered X-rays is measured by a semiconductor detector connected with a multichannel pulse-height analyser.

The relation between the modulus of the scatting vector k, the scatting angle 2ϑ and the energy E of the scattered photon is given by

$$k = \frac{4\pi}{\lambda} \sin \vartheta = \frac{4\pi}{hc} E \sin \vartheta = 1.014\ E \sin \vartheta\ [\text{Å}^{-1}\ \text{keV}];$$

h = Planck's constant, c = velocity of light.

In crystal diffraction the intensity of the scattered beam is peaked at discrete values of $k = \frac{2\pi}{d}$, where d is the interplanar spacing, characterized by indices hkl. The Bragg equation then reads as follows

$$Ed \sin \vartheta = \frac{hc}{2} = 6.199\ [\text{keV Å}].$$

All lines of the scattered spectrum are detected by use of a solid state detector with high energy resolution simultaneously.

Apparatus and Techniques

In energy-dispersive X-ray diffraction the sample is exposed to a white X-ray spectrum. We are using a standard 2 kW tungsten tube, operating at 60 kV and 33 mA. The X-ray tube produces polychromatic radiation in the energy range of 0 to 60 keV and good intensity results in the region of $\approx$ 15—45 keV. The sample is normally contained in a hole of diameter $\geqq$ 0.1 mm in a metallic gasket held between the diamond anvils and therefore it is necessary to collimate the incident beam so that it only incidents on the sample and not on the gasket. This has been achieved by us through the use of a glass capillary tube (ID = 0.5 mm) and a $\approx$ 1 mm thick plug of lead with a cylindrical hole of 0.1 mm. An intrinsic Ge-detector with an energy resolution of about 165 eV is mounted behind the diamond anvil cell to record data. We operate at a fixed scattering angle for $6° \leqq \vartheta \leqq 12°$. Variation of the diffraction angle causes an expansion or compression of the lines in the spectrum similar to a variation of λ in the conventional angular dispersion analysis.

The detector electronics include a high-voltage power supply, a preamplifier and a linear amplifier. The detector is combined with a multichannel analyser (VKA-4995) produced in the Centre for the Production of Scientific Instruments of the Academy of Science of the GDR. The multichannel analyser connects an analog detecting and analysing system with a digital unit and a microcomputer system (MPS 4944).

The spectra will be transferred from the multichannel analyser to the computer directly and a final data evalution is available during the time the next spectra is accumulated.

Spectra have been obtained in the qualitative phase analysis with counting times of typically 2000—8000 sec.

Fig. 4. Present set-up with the diamond anvil cell and the detector in front of the X-ray tube.

Results

A brief discussion of results on KBr will be given. The high pressure phase transformation in alkali halides from the NaCl-structure (B1) to the CsCl-structure (B2) has been repeatedly studied from the viewpoint of thermodynamic stability. Only a few investigations have been made on its kinetics. The knowledge of the transient behaviour of a transformation is of great importance in understanding its mechanism. In a reconstructive transformation, both the nucleation and the growth rate are small enough to permit their observation under appropriate external conditions. In a martensitic transformation, the rate of crystal growth is so great that direct observation of the growth stage is rather difficult under usual experimental conditions.

The mechanism of the B1—B2 transformation in alkali halides has not yet been established.

Features of a reconstructive transformation are reported by several authors. On the other hand at thermal conductivity studies on KBr we found some characteristics of a martensitic transformation. The B1—B2 transformation in KBr takes place at about 17.5 kbar. The energy pattern obtained from a polycrystalline sample of KBr by several pressures are represented in the Fig. 5.

The spectra were recorded at 2ϑ-diffraction angle of 13.8° and a counting time of 2000 sec.

Three diffraction peaks at energies of 15.7 keV, 22.2 keV and 27.0 keV corresponding

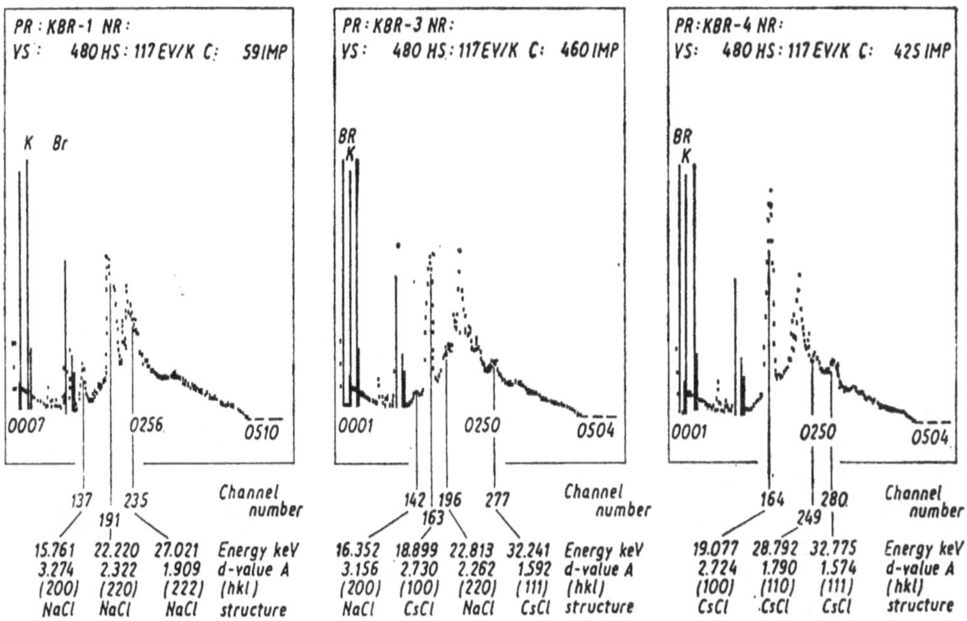

Fig. 5. Energy dispersive diffraction pattern of KBr at 0 GPa, 1.75 GPa and 2.6 GPa.

to d-spacings of 3.27 Å, 2.32 Å and 1.91 Å are seen in the first spectrum under normal pressure. These peaks are identified as (200), (220) and (222) of the NaCl-structure.

The second spectrum was taken at the phase transition point at 17.5 kbar. Both, the low pressure phase with the NaCl-structure and the high pressure phase with the CsCl-structure exist in the same sample.

The (200) and (220)-reflexes of the NaCl-structure disappear and the (100) and (111)-reflexes of the high pressure phase with the CsCl-structure are the main diffraction peaks.

Only the peaks of the CsCl-structure are seen in the third spectrum at a pressure of 26 kbar.

The phase transformation is concluded.

The first results obtained with this experimental arrangement show that the energy-dispersive X-ray diffraction under high pressure is a powerful tool for the detection of phase transitions and kinetic studies.

CHAPTER VIII

Planetary cosmogany and evolution

The growth of planets due to capture of solid grains by single planetary embryo

J. LELIWA-KOPYSTYŃSKI, Z. CZECHOWSKI

1. Introduction

The problem of the formation of terrestrial planets due to their accretion from the protosolar nebula material has not been solved fully unambiguously so far. All the existing models of accretion are based on some assumptions on the pre-accretional state of the protosolar nebula, such as shape and size, total mass, heavy elements to hydrogen-helium component mass ratio, temperature, pressure, condensational sequence, opacity and others. In numerical elaborations, these parameters are taken with large arbitrariness. In fact, the only one well-known feature of the Solar System in the accretion era are the post-accretional masses of planets: they were equal to their present masses. It is commonly assumed that in the formation period the orbits of planets were the same as the present ones. The latter, intuitively self-evident assumptions is, however, not well-documented, since the problem of the stability of the Solar System over the time interval elapsed from the end of accretion, i.e. $\sim 5 \cdot 10^9$ years, has not been solved.

Assuming the scheme of the accretion process in the form: dust $\rightarrow$ grains $\rightarrow$ planetesimals $\rightarrow$ planetary embryo $\rightarrow$ planet, the problem can be approached statistically (see e.g. NAKAGAWA, 1978; MATSUI and MIZUTANI, 1978; MATSUI, 1978 and 1979) appropriately for the swarm of planetesimals; or alternatively the growth of a single embryo can be considered (see e.g. SAFRONOV, 1969; MIZUTANI et al., 1972; WEIDENSCHILLING, 1974 and 1976; CORADINI et al., 1982 and 1983).

In the present paper, the latter approach is applied, in considering the growth of a single embryo as a result of sticking of solid grains moving round the proto-Sun on the Keplerian orbits. These orbits are contained in some feeding zones established for a given embryo. The present model is used both for terrestrial planets (from Mercury to Mars, excluding the Moon) and for the cores of giant planets (Jupiter's, Saturn's, Uranus' and Neptune's cores), about which it is assumed that in their compositions, and thus in their way of formation, they are also terrestrial-type bodies. Such an assumption about cores of giant planets was made for example by MIZUNO (1980), who investigated the stability of the hydrogen-helium envelope surrounding the Earth-like ($\varrho = 5.5\,\mathrm{g \cdot cm^{-3}}$) core growing at an accretion rate of $1 M_\oplus/10^6$ y, where $M_\oplus$ is the Earth's mass. GLADYSHEW and BUDTOW (1981) considered the accretion of hydrogen and helium on the giant planets cores which as it was supposed have been formed earlier.

The purpose of the present paper is to describe the growth of a planetary embryo by means of a model involving as few detailed assumptions as possible. This was achieved by introducing only assumptions on the shape of the feeding zone and the velocity distribution of grains in this zone. In this sense, the present papar is a development of the approach used by Mizutani et al. (1972) who analyzed the problem of the accretion of the Moon's embryo from a spherical cloud of grains. They received, however, very surprising results: the accretion time od the Moon was shorter than 10^3 years, and, for a particular case of initial numerical data, even shorter than one year.

Our calculations concerning the planetary accretion give an accretion time at least of the order of 10^7 years. Safronov and Vityazev (1985) in their review work inform that the planetary accretion time is frequently estimated to be $\sim 10^8$ years. They mention, however, that some models suggest the accretion a much shorter time even 10^5-10^4 years. The parameters of our model are: the thickness of the accretion layer H and two parameters of the velocity distribution v' and α. The parameter v' is estimated by formula (17); the only two "free" parameters are α and H.

2. Model of planetary embryo growth

It is assumed that the model describing the growth of the embryo as a result of its being bombarded by solid grains satisfies the following conditions:

1. The feeding zones are rings Ω_i with the thickness H_i and the radial expansion Δ_i, i.e.

$$\Omega_i = \left\{(x, y, z): a_i^2 \leqq x^2 + y^2 < (a_i + \Delta_i)^2 = a_{i+1}^2, -\frac{1}{2} H_i \leqq z \leqq \frac{1}{2} H_i\right\}. \tag{1}$$

The Sun lies at the point (0, 0, 0). The index "i" numbers the feeding zone of a successive embryo; unless there is some suspicion of ambiguity, it will be neglected further on.

2. The embryo, which is a sphere with the radius $R(t)$ and the constant density ϱ, moves at the velocity v_0 along a circular orbit contained in Ω.

3. The masses of all the grains are equal; their densities are equal to the density of the embryo, ϱ; the grain mass is much less than that of the embryo and big enough to neglect the gas drag forces.

4. The velocity distribution function of the grains, $f(\bar{x}, \bar{v}, t)$ does not depend on the position $\bar{x}$ and the time t in a given feeding zone, and it has the form of the Maxwell distribution function, truncated over the velocities $\bar{v}$:

$$f(\bar{v}) = \begin{cases} C \exp\left[-\frac{\alpha^2}{v'^2}(\bar{v} - \bar{v}_0)^2\right] & \text{for } |\bar{v} - \bar{v}_0| \leqq v', \\ 0 & \text{for other } \bar{v}, \end{cases} \tag{2}$$

where v' will be defined below. It follows from the normalization condition

$$C \int\limits_{|\bar{v}-\bar{v}_0| \leqq v'} \exp\left[-\frac{\alpha^2}{v'^2}(\bar{v} - \bar{v}_0)^2\right] d\bar{v} = 1, \tag{3}$$

that the constant C is

$$C = \frac{1}{2\pi} \frac{\alpha^3}{v'^3} \left[\frac{1}{2} \sqrt{\pi}\, \Phi(\alpha) - \alpha \exp(-\alpha^2)\right]^{-1}, \tag{4}$$

where $\Phi(\alpha)$ is the probability integral

$$\Phi(\alpha) = \frac{2}{\sqrt{\pi}} \int_0^{\alpha} \exp(-y^2) \, dy. \tag{5}$$

5. Each feeding zone is a closed region, i.e. the grains do not dissipate outside nor do they come from outside. The interactions among the grains do not bring about changes in their number or mass of any of them, nor in the velocity distribution function. All the grains impacting the embryo join with it, thus the sticking coefficient is assumed to be one.[1]

The assumption 1. and 5. do not define the real dimensions of zone of gravitational influence of an embryo under consideration but only fix the dimension and mass (disposable for accretion) of an appropriate feeding zone.

According to the above assumptions, the increase in the mass of the embryo in unit time, as a result of its being bombarded by grains at different velocities, is

$$\frac{dM}{dt} = \frac{M_0 - M(t)}{V} \int\limits_{|\bar{v}-\bar{v}_0| \leq v'} S_{eff}[M(t), R(t), |\bar{v} - \bar{v}_0|] \, |\bar{v} - \bar{v}_0| \, f(\bar{v}) \, d\bar{v}, \tag{6}$$

where

$$S_{eff}[M(t), R(t), |\bar{v} - \bar{v}_0|] = \pi R^2(t) \left[1 + \frac{2GM(t)}{R(t)\,(\bar{v} - \bar{v}_0)^2}\right]. \tag{7}$$

S_{eff} is the effective cross-section for coagulation of grains with the embryo with the mass $M(t)$ and the radius $R(t)$, when the grain velocity far from the embryo is $\bar{v}$. The other quantities in formula (6) are:

- the gravitation constant $G = 6.672 \cdot 10^{-11}\ \mathrm{N \cdot m^2 \cdot kg^{-2}}$,
- the initial mass of the grain cloud, M_0, equal to the final mass of the embryo, i.e. to the present mass of the terrestrial planet or of the terrestrial-like core of the giant planet,
- the volume of the feeding zone

$$V = \pi H[(a + \Delta)^2 - a^2], \tag{8}$$

where H is the mean thickness of the ring-like feeding zone,
- the term $[M_0 - M(t)]/V$ is the density of the grain cloud.

Substitution of $f(\bar{v})$ in the form of (2) into equation (6) and application of the formulae

$$M(t) = \frac{4}{3} \pi \varrho R^3(t), \qquad M_0 = \frac{4}{3} \pi \varrho R_0^3 \tag{9}$$

[1] Such an assumption, although commonly used, is, however, incorrect for the initial stage of accretion, when the gravitational component of the effective cross-section is small. Experimental papers (e.g. Hartmann, 1978, Leliwa-Kopystyński et al., 1984) indicate the lack of the sticking effect in rock-rock collision at room temperature, or that the result of the impact (sticking, partial sticking, rebound, catastrophic destruction) depends significantly on the physical conditions characterizing the impact.

give after integration over $\bar{v}$ a differential equation describing time changes in the dimensionless radius of the growing embryo:

$$\frac{\mathrm{d}x}{\mathrm{d}t} = \frac{1}{H} A^2 p^2 (1 - x^3)(1 + B^2 q^2 x^2), \tag{10}$$

where R_0 is the final radius of the embryo and

$$x(t) = \frac{R(t)}{R_0} \tag{11}$$

is its dimensionless radius. The coefficients

$$A_2 = \frac{2}{3\sqrt{\pi}} \frac{R_0^2 v'}{(a + \Delta)^2 - a^2}, \tag{12}$$

$$B^2 = \frac{8}{3} \pi G \frac{\varrho R_0^2}{v'^2}, \tag{13}$$

$$p^2 = \frac{1}{\alpha} \frac{1 - (\alpha^2 + 1) \exp(-\alpha^2)}{\Phi(\alpha) - \frac{2}{\sqrt{\pi}} \alpha \exp(-\alpha^2)}, \tag{14}$$

$$q^2 = \alpha^2 \frac{1 - \exp(-\alpha^2)}{1 - (\alpha^2 + 1) \exp(-\alpha^2)} \tag{15}$$

depend on well-established parameters characterizing the feeding zone (see below and Table 1) and on the so far arbitrary parameter α characterizing the velocity distribution.

When $v' \to \infty$ and $\alpha \to \infty$ in such a way that $\alpha/v' = \text{const.}$ and on the assumption that the feeding zone has the shape of a sphere (which leads to a slightly different definition of the coefficient A^2), equation (10) is reduced to the one consideed by MIZUTANI et al. (1972) and used for the problem of the growth of the Moon's embryo. It is also possible to compare the equation (10) with the equation of SAFRONOV (1969) having the form

$$\frac{\mathrm{d}x}{\mathrm{d}t} \sim \text{initial surface density} \cdot (1 - x^3)(1 + 2\theta).$$

In this equation the constant term $(1 + 2\theta)$ is introduced instead of the term $(1 + B^2q^2x^2)$ in equation (10); θ is the Safronov's parameter being $\theta = 3$ for gas free accumulation. The thickness of the feeding zone in Safronov's equation is no more an important parameter and the feeding zone parameter is the surface density.

The parameters a, Δ, v', for given a feeding zone, occuring in formulae (12)—(15) will be defined in the following way (see Table 1):

— The parameter a_i for the i-th feeding zone is equal (according to KUSAKA et al., 1970) to the geometrical mean of the mean solar distances (designated as a_i) of two neighbouring present planets, i.e.

$$a_i = (\bar{a}_{1-1} \bar{a}_{i+1})^{1/2}, \qquad \Delta_i = a_{i+1} - a_i. \tag{16}$$

Table 1. Numerical data as used in our model: a — inner radius of the feeding zone, Δ — the width of the feeding zone, v' — the velocity defined by formula (17), ϱ and R_0 the average density and radius, respectively, of the present planet, or of the present core of a giant planet. The mass ranges of the cores of giant planets are fitted to many models concerning the structure of giant planets. This table also gives a comparison of the solar distances $\bar{a}$ of the present planets and the solar distances a_0 of the growing embryos in our model. The values of a and Δ have been taken from Kusaka et al. (1970) (formula (16)). $M_\oplus$ and $\varrho_\oplus$ are the Earth's mass and average density, respectively.

Planet	a (a.u.)	$a+\Delta$ (a.u.)	v' (km $\times s^{-1}$)	ϱ (10^3 $\times$kg m^{-3})	R_0 (10^6 m)	$\bar{a}$ (a.u.)	a_0 (a.u.)	Remarks: ϱ and R_0 calculated according to the following assumptions
Mercury	0.30	0.53	6.7	5.460	2.440	0.387	0.391	
Venus	0.53	0.85	4.3	5.250	6.055	0.732	0.662	
Earth	0.85	1.23	2.7	5.520	6.371	1.000	1.014	
Mars	1.23	2.06	3.1	3.930	3.389	1.524	1.566	
Jupiter's core	3.82	7.04	2.0	16.560	11.990–15.106	5.203	5.070	$M_{Jup.core}/M_\oplus = 20-40$, $\varrho_{Jup.core}/\varrho_\oplus = 3$
Saturn's core	7.04	13.5	1.6	11.040	9.188–11.576	9.539	9.520	$M_{Sat.core}/M_\oplus = 6-12$, $\varrho_{Sat.core}/\varrho_\oplus = 2$
Uranus' core	13.5	24.0	1.0	5.520	6.371–8.027	19.18	17.63	$M_{Uran.core}/M_\oplus = 1-2$, $\varrho_{Uran.core}/\varrho_\oplus = 1$
Neptune's core	24.0	38.1	0.63	5.520	6.371–8.027	30.06	30.06	$M_{Nept.core}/M_\oplus = 1-2$, $\varrho_{Nept.core}/_\oplus\varrho = 1$

Such a definition involves some arbitrariness in determining a_i for a planet farthest from the Sun and that closest to it. After Kusaka et al. (1970), it is assumed that $a_1 = 0.30$ a.u. (for Mercury).

— The parameter v' is defined as half the difference between the orbital velocities on the orbit with the radius a and on that with the radius $a + \Delta$:

$$v' = \frac{1}{2}\,[v(a) - v(a + \Delta)] = \frac{1}{2}\,(GM_\odot)^{1/2}[a^{-1/2} - (a + \Delta)^{-1/2}], \tag{17}$$

where the mass of the Sun $M_\odot = 2 \cdot 10^{30}$ kg. The orbital velocity v_0 and the radius of the orbit a_0 of the embryo are then defined by the formulae

$$v_0 = v(a) - v' = \frac{1}{2}\,[v(a) + v(a + \Delta)], \tag{18}$$

$$a_0 = \frac{GM_\odot}{{v_0}^2}. \tag{19}$$

It can be seen from formulae (18), (19) and from Table 1 that the values of a_0 and v_0 determining the solar distance and the orbital velocity of the embryo by means of definition (16) are close to the corresponding values of $\bar{a}$ and $\bar{v}_0 = (GM_\odot/\bar{a}^{1/2}$ for the present planets. The velocities v' calculated from formula (18), see Table 1, limiting the embryo/grain relative velocity range are of the same order as those presented by Wetherill (1978).

As far as the parameter of the velocity distribution α is concerned, no estimations of the variability range of this parameter are performed here. It can only be noted that in the boundary case $\alpha \to 0$ the velocity distribution converges to the uniform one. For high values of α, all grains orbit at a velocity close to that of the embryo. The effect of changes in the parameter α on the accretion process will be discussed in the next section.

3. Numerical results

We shall discuss solutions of the accretion equation (10) depending on the parameters occurring in it and on the different geometrical models of accretion disk which determine the thickness of the feeding zone, H, of the planet under study. This equation will be used to describe the growth of terrestrial planets and the growth of the cores of giant planets. As to the data on the mass cores of giant planets, cautions estimates of the lower and upper limits of these masses — see Table 1 — will be assumed. If one were to assume larger core masses, with the other parameters unchanged, a shorter accretion time would occur. The accretion time t_a is understood to be the time necessary for the radius of a planet or the core of a giant planet to reach 98% of the value of the final radius R_0. Such a definition was used by Mizutani et al. (1972), who applied equation (10) to the problem of the accretion of the Moon. At constant density of the growing planet, the value of $0.98R_0$ corresponds to the accumulation of 94% of the final mass of the planet, M_0.

The equation (10) is integrated within the interval between the initial radius and that at the time t, i.e. in dimensionless variable from $x(0) \equiv x_e = R_e/R_0$ to $x(t) = R(t)/R_0$ (R_e denotes the radius of the embryo and R_0 is the final radius of the present terrestrial planet or the corse of the present giant planet). The dimensionless integral can be expressed by elementary functions. The thickness of the feeding zone, H, can depend on the distance of the feeding zone from the Sun.

The accretion of the terrestrial planets (Mercury, Venus, Earth, Mars) and of the cores of giant planets (Jupiter, Saturn, Uranus, Neptune) will be analysed. Table 1 gives a list of input data used to calculate the coefficients A^2 and B^2 as definad by formulae (12) and (13). For comparison, data are also given on the mean solar distances of the present planets, $\tilde{a}$, and those of the growing embyos, a_0. On the assumption that the data in Table 1 are known with good accuracy, the question of the accretional growth of the planetary embryo in the present model becomes a two-parameter problem[2], which depends on the parameter of the velocity distribution, α, and on the thickness H of the feeding zone. The effect of these factors on the process of accretion will now be studied.

---

[2] In reality there is still another parameter, namely the radius of the embryo, R_e, or equivalently the dimensionless quantity x_e, see formula (20). However, the effect of this parameter on the accretion process is negligible only if $x_e < 0.01$. In the present calculations, it was assumed that always $R_e = 10$ km, and therefore, $x_e \leqq 10/R_☿ = 0.0041$, where $R_☿$ is the radius of Mercury.

The influence of the parameter α will first be considered. In order to do so, it will be assumed that the thickness of the feeding zone is constant. Fig. 1 shows the dependence of the dimensionless radius of the growing Earth on time for different values of α. Changes in the parameter α hardly affect the accretion time, but exert a significant influence on the accretion rate: an increase in α causes the growth curve of the radius of the planetary embryo to become steeper and steeper in the main period of accretion. The curves of accretion shown in Fig. 1 allow the value of the parameter $\alpha = 2$ to be

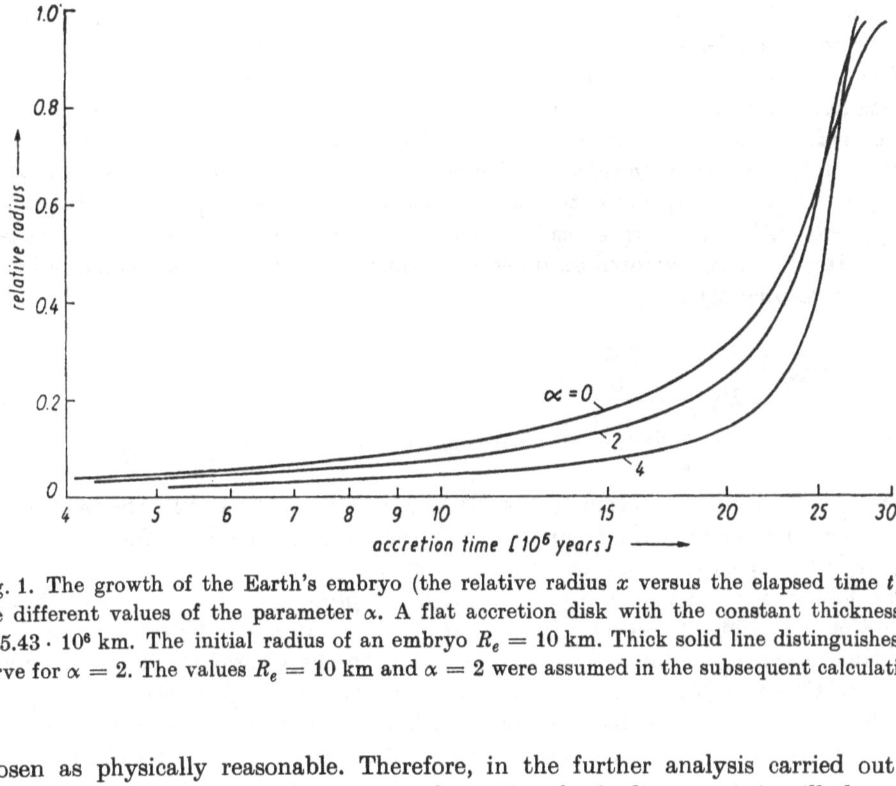

Fig. 1. The growth of the Earth's embryo (the relative radius x versus the elapsed time t) for the different values of the parameter α. A flat accretion disk with the constant thickness $H_0 = 5.43 \cdot 10^6$ km. The initial radius of an embryo $R_e = 10$ km. Thick solid line distinguishes the curve for $\alpha = 2$. The values $R_e = 10$ km and $\alpha = 2$ were assumed in the subsequent calculations.

chosen as physically reasonable. Therefore, in the further analysis carried out for changes in the thickness of the accretion layer (i.e. the feeding zone), it will always be assumed that $\alpha = 2$.

It is clear from the equation (10) that the accretion time is proportional to the thickness of the accretion layer H. Therefore the choice of the geometrical model of the shape of the accretion layer is essential for the analysis of the accretion process. The three models of the accretion disk are considered:

(a) A flat accretion disk with constant thickness

$$H = H_0 = 5.43 \cdot 10^6 \text{ km}. \tag{20}$$

The value of H_0 was so selected that this model (a) could be the same as models (b) and (c) for the accretion of the embryo of the Earth. This value corresponds to the model of protoplanetary nebula proposed by Cameron and Pine (1973). For comparison we mention that Goldreich and Ward (1973) proposed the thickness of the gaseous disk in the vicinity of the Earth's orbit as $2 \cdot 10^7$ km. According to different works of SAFRONOV this thickness is of the order of 10^7 km.

(b) An accretion disk with thickness increasing linearly as the distance from the Sun increases. For the respective feeding zone

$$H = 2\left(a + \frac{\Delta}{2}\right) \tan 1^\circ. \tag{21}$$

This model of the shape of the accretion disk, together with an angle of 2° assumed between the limiting surfaces of the disk, corresponds roughly, up to the orbit of Uranus, to the model of protoplanetary nebula proposed by CAMERON and PINE (1973). In fact the model of the accretion disk with thickness increasing as solar distance increases have been used also by SAFRONOV (1969) (see equation (5) of Chapter 3 which led to his fundamental equation of the accretion, equation (14) of Chapter 9).

(c) An accretion disk with thickness decreasing as the distance from the Sun increases. CORADINI et al. (1980) presented such a shape of low-mass protoplanetary cloud ($M_{\text{cloud}} = 0.01 M_\oplus$) as being a result of settling of solid grains dissolved in a gaseous nebula. Here we have adopted an inversely proportional relation and therefore for the respective feeding zone

$$H = H_0 \frac{a_\oplus + \frac{\Delta_\oplus}{2}}{a + \frac{\Delta}{2}} \tag{22}$$

where $a_\oplus$ and $\Delta_\oplus$ are the parameters of the feeding zone of the Earth. The relation (22) is obviously not valid if the feeding zone is very near to the Sun. Therefore an unexpected long formation time results for Mercury.

The solutions $x = x(t)$ of the equation (10) are presented on the Figs. 2a, b, c for models (a), (b) and (c) respectively. The accretion of Mercury, Venus and the Earth calculated according to models (a) and (b) leads to the generally accepted accretion time $t_a \lesssim 10^8$ years (Table 2). For Mars and the cores of the giant planets, models (a) and (b), Figs. 2a, b and Table 2, show a much longer accretion time (a time which is even longer than the existence time of the protosolar system proposed by a large number

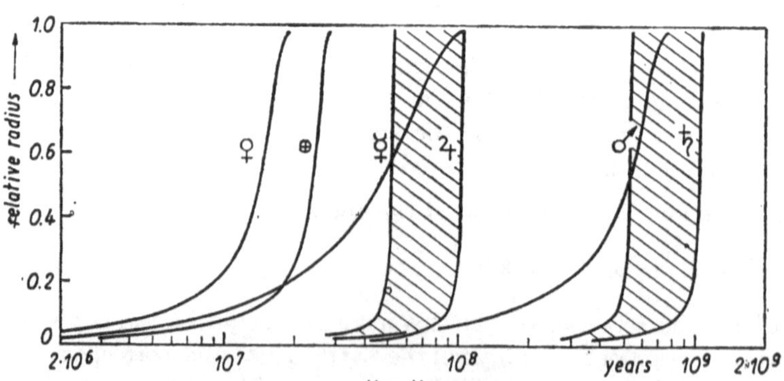

Fig. 2 a

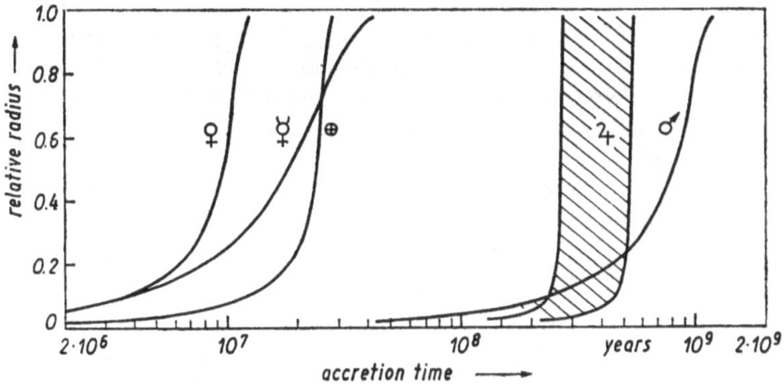

Fig. 2b

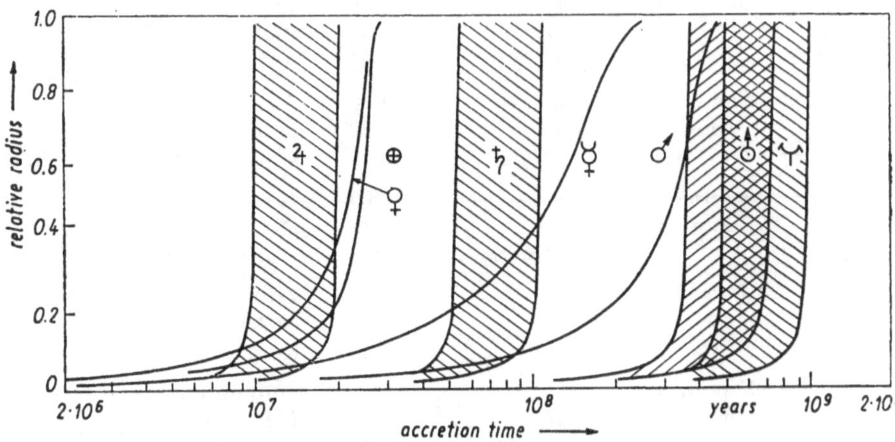

Fig. 2c

Fig. 2. The growth of planets: Mercury ☿, Venus ♀, Earth ⊕, Mars ♂ and the cores of Jupiter ♃, Saturn ♄, Uranus ⛢ and Neptune ♆ for the different geometrical models of the protoplanetary nebula. $H_0 = 5.43 \cdot 10^6$ km, $R_\iota = 10$ km, $\alpha = 2$. The shaded areas present the growth of cores of giant planets: left-side boundary represents the high-mass core and the right-side the low-mass core. The figure could be treated as an illustration of time sequence of formation of planets only under assumption that the starting moment was the same for all planets and cores.

a) A flat accretion disk (formula (20)). Uranus and Neptune are beyond the right side of this figure.

b) An accretion disk with its thickness increasing in the direct proportion to the solar distance (formula (21)). Saturn, Uranus and Neptune are beyond the right side of this figure.

c) An accretion disk with thickness decreasing as the distance from Sun increases (formula (22)).

Table 2. The elapsed accretion time t_a in 10^8 years, the accretion layer thickness H in 10^6 km, and the maximum value of the accretion rate $(dR/dt)_{max}$ in cm · y^{-1}, as found for models (a)–(c) of the accretion layer. The used values of the parameters: $H_0 = 5.43 \cdot 10^6$ km, $R_e = 10$ km, $\alpha = 2$; the lower limit of radius R_0 of cores of giant planets is used.

Planet		*a*	*b*	*c*
Mercury	t_a	1.06	0.42	2.65
	H	5.43	2.17	13.6
	$(dR/dt)_{max}$	3.04	7.62	1.21
Venus	t_a	0.19	0.13	0.28
	H	5.43	3.60	8.18
	$(dR/dt)_{max}$	88.3	133	58.6
Earth	t_a	0.28	0.28	0.28
	H	5.43	5.43	5.43
	$(dR/dt)_{max}$	101	101	101
Mars	t_a	7.65	12.1	4.84
	H	5.43	8.59	3.43
	$(dR/dt)_{max}$	0.90	0.57	1.42
Jupiter's core	t_a	1.06	5.55	0.20
	H	5.43	28.3	1.04
	$(dR/dt)_{max}$	214	41	1116
Saturn's core	t_a	10.9	108	1.11
	H	5.43	53.6	0.55
	$(dR/dt)_{max}$	12.9	1.30	1269
Uranus' core	t_a	138	2491	7.66
	H	5.43	97.9	0.30
	$(dR/dt)_{max}$	0.53	0.029	9.61
Neptune's core	t_a	303	9073	10.1
	H	5.43	162	0.18
	$(dR/dt)_{max}$	0.39	0.013	11.5

of authors): the respective accretion curves for the cores of Saturn, Uranus and Neptune run to the right beyond the time interval included in Figs. 2a, b.

A shape of accretion disk which can assure the physically reasonable accretion time can be presented by a model with thickness decreasing as the solar distance increases, for example model (c). Fig. 2c presents the respective accretion curves; see also Table 2 presenting the appropriate accretion times.

Changes in the accretion rate dR/dt, versus the relative radius x and versus $(t - t_a)$, i.e. versus time measured with respect to the moment t_a, are presented, respectively, in Figs. 3 and 4.

Fig. 3 shows the accretion rate versus the relative radius for all planets or cores considered, calculated for model (a). It can be seen that the accretion rates of Mars and Mercury during the main stage of accretion are lower by two orders of magnitude than those for Venus and the Earth. A similar regularity also occurs in models (b) and (c), as is indicated by values of $(dR/dt)_{max}$ listed in Table 2. As far as the cores of the giant

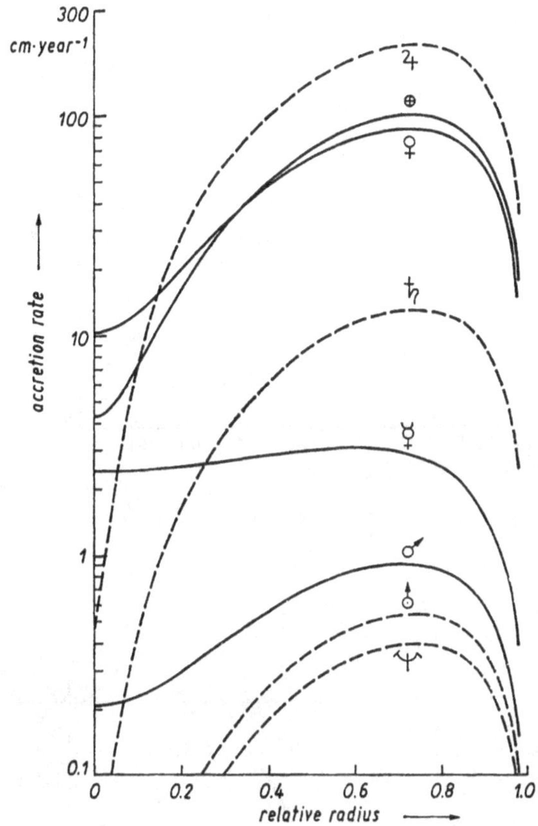

Fig. 3. The accretion rate dR/dt, in cm · y^{-1} versus the relative radius x, according to model (a). The parameters take the same values as in Fig. 2a. The lower limit of radius R_0 of cores of giant planets is used.

planets are concerned, it was found that $(dR/dt)_{max}$ is for Jupiter's core least sensitive to a change from one accretion layer model to another (see Fig. 4b, Table 2). For the embryos of the other outer planets the selection of a particular model is of essential significance.

4. Conclusions

The model of the growth of a single (in an appropriate feeding zone) planetary embryo considered in this paper, cannot be used until some moment of the evolution of protoplanetary bodies: at the initial stage, there is a large number of planetesimals with similar masses and there is no reason to distinguish any of them as the future planetary embryo, on the other hand, the final result of accretion was the formation of the present planets whose orbits do not intersect[3]. Thus, the planetary embryo orbiting in their

[3] Excluding the case of Neptune's and Pluto's orbits, and probably that of the Earth-Moon double planet.

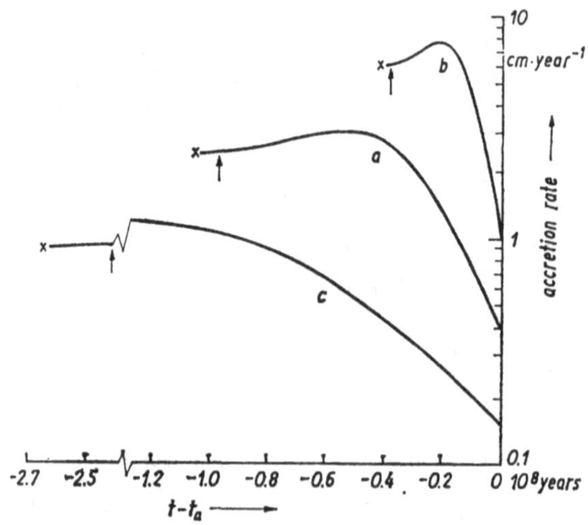

Fig. 4a

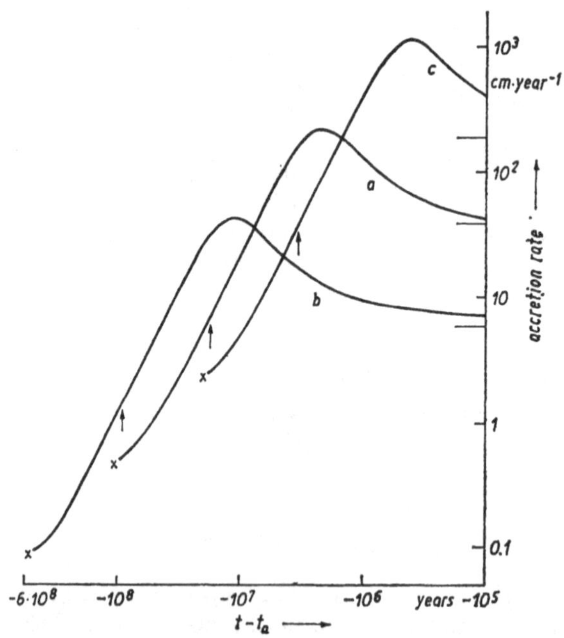

Fig. 4b

Fig. 4. The accretion rate $\mathrm{d}R/\mathrm{d}t$, in cm · y^{-1} versus time $(t - t_a)$, in 10^8 years, according to models (a)—(c). The parameters take values as in Fig. 2. The arrows denote such points on the respective curves to which there corresponds the relative radius of growing embryo, equal to 0.1 of the final radius. The crosses denote the initiation of accretion. (a) Mercury, (b) Jupiter's core, where $M_{\mathrm{Jup.core}}/M_{\oplus} = 20$.

own feeding zones, filled by bodies much less than itself, must have existed. In this sense, each such embryo can be treated as a single embryo. The question of the moment when a single embryo, greater than the other planetesimals, was formed is not considered in this paper. In this sense, the assumption that at the time $t = 0$ there is an embryo with a set radius $R_e = 10$ km is convenient but arbitrary. Any improvement of the model considered here is related to the neglecting or weakening of assumptions 1–5 given in Section 2. This involves the necessity of introducing additional parameters.

It can be concluded from the numerical results given in Section 3 that:

1. The model of the growth of a single embryo for all three geometrical cases considered here leads to physically reasonable values of the accretion time for planets from Mercury to Jupiter's core. However only such a feeding zone for which the thickness decreases as the distance from the Sun increases can be applied for the accretion of all planets.

2. It was found that the accretion times are as follows: Mercury (42–265), Venus (13–28), Earth (28), Mars (480–1200), Jupiter's core (10–550), Saturn's core (>55), Uranus' core (>380), and Neptune's core (>500). In the brackets the accretion times are given in million years, calculated for different geometrical models of feeding zone. The unusual long accretion time of Mars can be interpreted as an upper limit; the assumption concerning a closed feeding zone (ass. 5) is probably violated in the case of Mars because of the strong influence of Jupiter. It was also found that for physically acceptable geometry of the accretion layer the main stage of accretion from about 10% to 98% of the present radius of the planets or planetary cores takes 2 to 100 million years. An exception is Mars, for which this time interval in models (a), (b) and (c) is, respectively, 609, 964 and 380 million years.

3. On the basis of the curves of the accretion rate, dR/dt, shown in Fig. 3 and from the data on the accretion time given by Table 2, one can attempt to explain the difference in the cratering of the surfaces of terrestrial planets. As can be seen, small terrestrial planets, namely Mercury and Mars[4], grew slowly: over the most part of the accretion process, in particular at its final stage determining the relief of the surface, their accretion rate was lower by at least an order of magnitude than the accretion rates of Venus and the Earth. As the embryos of terrestrial planets grew, planetesimals emerged in their feeding zones (in reality and not according to the model of a single embryo). The bombardment by these planetesimals was the reason for the formation of the strongly cratered surface of Mars, and particulary that of Mercury. Large terrestrial planets, Venus and the Earth, grew much faster and their masses increased at the expense of small solid grains from their feeding zones rather than at the expense of planetesimal-type bodies, which did not form in such a number as that of Mercury's or Mars's zones. Therefore, the cratering ("big bombardment") occured in a much weaker way for Venus and for the Earth.

---

[4] This probably also applies to the Moon provided only its accretion occured in a feeding zone round the Sun and not round the Earth.

References

Cameron, A. G. W.; Pine, M. R.: Numerical models of the primitive solar nebula. Icarus **18** (1973) 377–406.

Coradini, A.; Magni, G.; Federico, C.: Time evolution of grains in the protosolar nebula. The Moon and the Planets **22** (1979) 47–61.

Coradini, A.; Federico, C.; Magni, G.: Ganymede and Callisto: Accumulation heat content. In: The comparative study of the planets (A. Coradini and M. Fulchignoni, Eds.). D. Reidel Publ. Comp. 1982, p. 61–70.

Coradini, A.; Federico, C.; Lanciano, P.: Earth and Mars: early thermal profiles. Phys. Earth Planet Interiors **31** (1983) 145–160.

Gladyshev, G. P.; Budtov, V. P.: The Solar System evolution. The Moon and the Planets **25** (1981) 413–425.

Goldreich, P.; Ward, W. R.: The formation of the planetesimals. The Astronomical Journal **183** (1973) 1051–1061.

Hartmann, W. K.: Planet formation: Mechanism of early growth. Icarus **33** (1978) 50–61.

Kusaka, T.; Nakano, T.; Hayashi, C.: Growth of solid particles in the primordial solar nebula. Progr. Theor. Phys. **44** (1970) 1580–1595.

Leliwa-Kopystyński, J.; Taniguchi, T.; Kondo, K.; Sawaoka, A.: Sticking in moderate velocity oblique impact: Applications to planetology. Icarus **57** (1984) 280–293.

Matsui, T.: Collisional evolution of mass-distribution spectrum of planetesimals. Proc. Lunar Planet. Sci. Conf. 9th (1978) **1**–13.

Matsui, T.: Collisional evolution of mass-distribution spectrum of planetesimals II. Proc. Lunar Planet. Sci. Conf. 10th (1979) 1881–1895.

Matsui, T.; Mizutani, H.: Gravitational N-body problem on the accretion process of terrestrial planets. Icarus **34** (1978) 146–172.

Mizuno, H.: Formation of the giant planets. Progr. Theor. Phys. **64** (1980) 544–557.

Mizutani, H.; Matsui, T.; Takeuchi, H.: Accretion process of the Moon. The Moon **4** (1972) 658–671.

Nakagawa, Y.: Statistical behaviour of planetesimals in the primitive solar system. Progr. Theor. Phys. **58** (1978) 1834–1851.

Safronov, V. S.: Evolution of the Protoplanetary Cloud and Formation of the Earth and the Planets. Nauka, Moscow 1969. (English translation NASA TT F-677, 1972).

Safronov, V. S.; Vityazev, A. V.: Origin of the Solar System. Sov. Sci. Rev. E Astrophys. Space Phys., vol. 4, 1985, 1–98.

Weidenschilling, S. J.: A model for accretion of the terrestrial planets. Icarus **22** (1974) 426 to 435.

Weidenschilling, S. J.: Accretion of the terrestrial planets II. Icarus **27** (1976) 161–170.

Wetherill, G. W.: Accumulation of the terrestrial planets. In: Protostar and Planets (T. Gehrels, Ed.) (1978) 565–598.

On the evolution of the earth's core

S. Franck

Introduction

At the present time it is generally accepted that the planets have been formed by the accumulation of solid particles and bodies (planetesimals). This hypothesis was brought into planetary cosmogony explicitly and clearly by the Soviet scientist O. Yu. Schmidt (see e.g. Schmidt, 1959). Details are described in the book "Evolution of the Preplanetary Cloud and the Formation of the Earth and Planets" by Safronov (1969). This process may be divided into two stages: a) evolution of the protoplanetary gas-dust cloud rotating around the Sun and its transformation into a swarm of solid bodies, b) accumulation of bodies and the growth of the biggest ones ro form the planets. In this way, the planet Earth is formed by homogeneous accretion as shown in Fig. 1 and

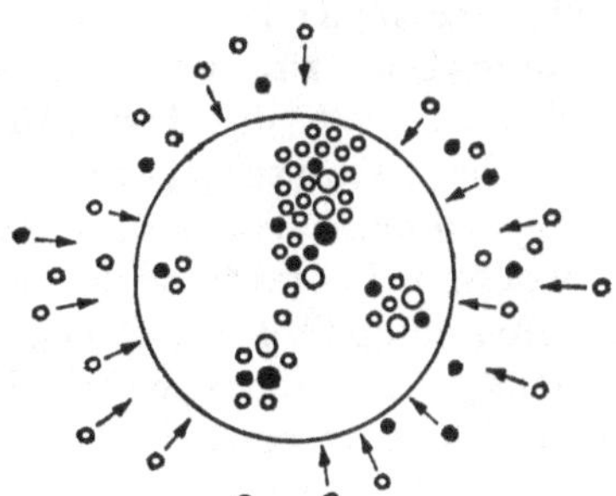

Fig. 1. Homogeneous accretion of the Earth.

the interior of the proto-Earth shows a homogeneous distribution of silicate- and iron-rich planetary matter. The following process of differentiation cepends on the temperature distribution on the Earth because it is very important whether any component of planetary matter is liquid or solid at the corresponding pressure and what is the viscosity of the unmelted material. According to Elsasser (1963), at the beginning of core formation great blobs of iron (that is iron compounds as Fe—S) originate from a radius of some hundred kilometers. Because of a Rayleigh-Taylor-instability these blobs sink through the mantle and cause core formation (see Fig. 2). In this way, at more than $4 \cdot 10^9$ y (4 Ga) ago the whole Earth's core was in a liquid state, but after that the slow cooling of the entire Earth led to freezing of core material causing formation of a solid inner core, as shown in Fig. 3. In the following chapters we want to discuss some details of this process.

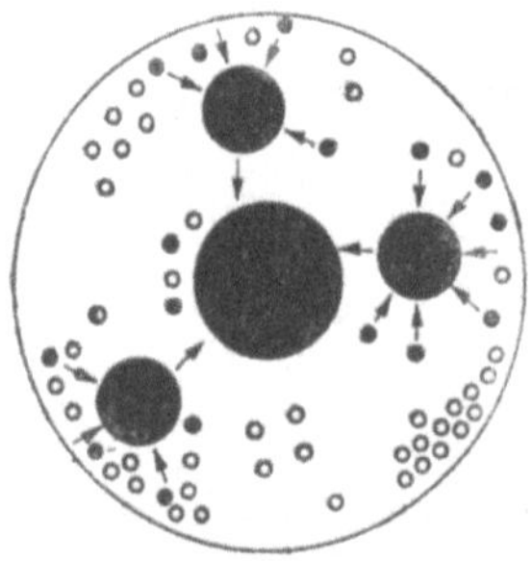

Fig. 2. Formation of a liquid Earth's core.

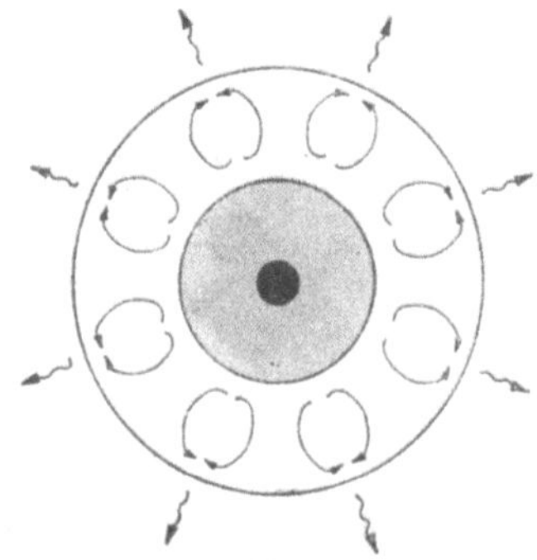

Fig. 3. Whole planet cooling and formation of the inner core.

Dynamics and energetics of the outer core

In the following we present a model of the dynamics in the outer core, first proposed in this form by Franck (1982): The core material is an alloy of iron and sulphur. The corresponding phase diagram of the system Fe—S at high pressures and temperatures shows the existence of a liquid immiscibility field between the Fe-endmember and the entectic. Above the liquidus curve there exist two liquids: a heavy liquid with a low sulphur content of 2.6 wt.-% S and a lighter liquid with 15 wt.-% A. If the solid inner core freezes from such an allow it has a smaller mass fraction of light material than the outer core. Referring to our proposed phase diagram (Franck, 1982), we conclude that the inner core is pure metal because it freezes from the heavy liquid with only about 2.6 wt.-% S. This property requires an upward flux of sulphur at the inner core boundary. Because of liquid immiscibility above the inner core boundary, the sulphur may not alloy with the heavy liquid. The only possibility is nucleation of liquid droplets with sulphur content of about 15 wt.-% (see Fig. 4).

The permanent flux of sulphur causes the droplets to grow up to such a size where bouyancy is strong enough to overcome drag forces. It was computed by Franck (1982) that these droplets are also stable because of the stability condition $R \ll 1$ m.

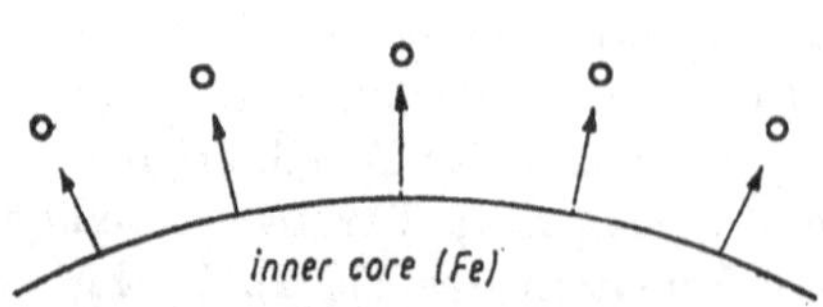

Fig. 4. Flux of sulphur and nucleation of droplets near inner-core-boundary.

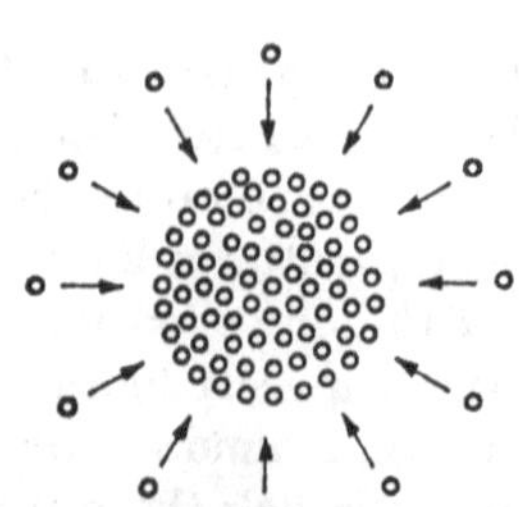

Fig. 5. Coagulation of droplets to form aggregates.

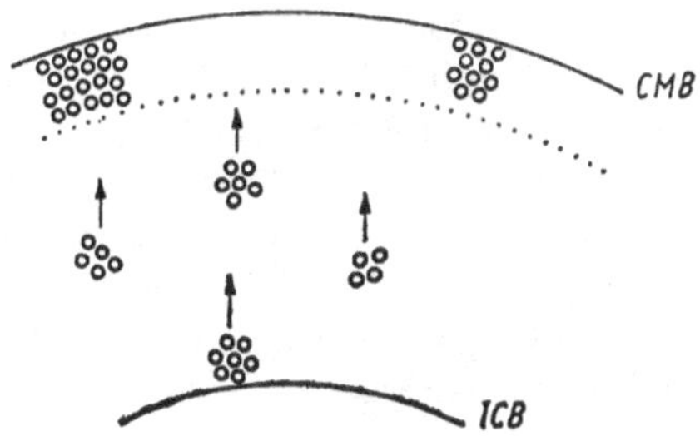

Fig. 6. Ascend of aggregates from inner-core-boundary (ICB) to form a stably stratified layer near core-mantle-boundary (CMB).

Furthermore, we propose a coagulation of droplets to form aggregates of a characteristic dimension $L > 10^4$ m (see Fig. 5). This condition follows from theory of hydrodynamic dynamo action.

The transport of light material (aggregates of droplets) (see Fig. 6) leads to a rearrangement of core material connected with a release of gravitational potential energy at a rate of about $2.26 \cdot 10^{11}$ W. This power is capable of driving a dynamo. A rouhg estimation of the size of the magnetic field can be found by equating this power to the ohmic power dissipation and using a formula given by LOPER and ROBERTS (1981). The result is an average toroidal magnetic field of about $1.5 \cdot 10^{-2}$ (or 150 G). Such a field can be classified as one of mean size.

The present state of the Earth's core

Taking into account the discussion in the foregoing chapter we can come to the following conclusions about the present state of the Earth's core:

The inner core consists of pure iron, above the inner-core-boundary (ICB) there is the heavy Fe-S-liquid with about 2.6 wt.-% S and at the top of the outer core the lighter Fe-S-liquid with about 15 wt.-% S is stably stratified. This is sketched in Fig. 7.

If the present inner core with a mass of $9.7 \cdot 10^{25}$ g iron has grown from the heavy

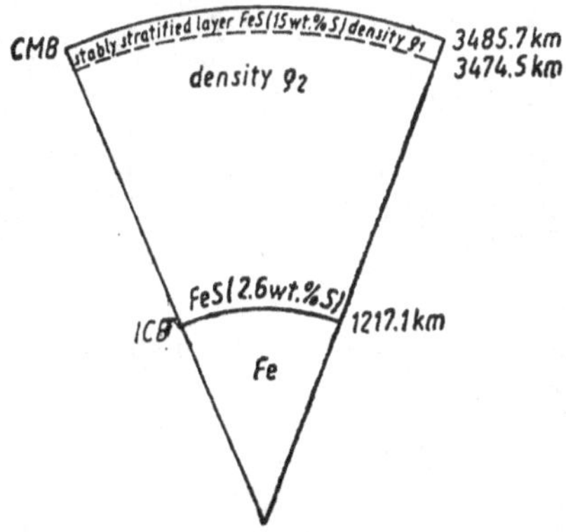

Fig. 7. Present structure of the Earth's core.

Fe-S-liquid at the ICB, then this process releases about $2.6 \cdot 10^{24}$ g sulphur which forms about $1.7 \cdot 10^{25}$ g light Fe-S-liquid. With a density of about 10 g/cm³ this light Fe-S-liquid occupies a volume of $1.7 \cdot 10^{9}$ km³ which corresponds to a spherical shell growing from CMB to a depth of about 11 km. The density contrast between the stably stratified layer and the rest of the outer core may be roughly estimated from an extrapolation of experimental results (Ahrens, 1979) to conditions as near CMB. In this way we found for the density of light Fe-S-liquid $\varrho_1 = 10/\text{cm}^3$ while the density of the heavy Fe-S-liquid should be in the range $\varrho_2 = 10-11$ g/cm³.

Reflection of seismic waves

According to the results of the preceding chapter the density contrast between the stably stratified layer and the rest of the outer core should cause a reflection of seismic waves near CMB. For a quantitative estimation we use the formula of the reflection coefficient R for rays incident from below at near-normal incidence (Lees et al. 1983):

$$R = \sinh\left[\frac{2\pi^2}{m\lambda}\left(\frac{I_2}{I_1} - 1\right)\right]\left\{\sinh\left[\frac{2\pi^2}{m\lambda}\left(\frac{I_2}{I_1} + 1\right)\right]\right\}^{-1}, \tag{1}$$

where λ is the wavelength of the incident ray; $I_1 = \varrho_1 v_{P1}$ and $I_2 = \varrho_2 v_{P2}$ are the acoustic impedances above and below the discontinuity. For $m = 10/L$, about nine tenth of the material property changes occur over the transition width L. The quantity L maybe also considered as the roughness of this interface between the heavy and light liquid. This interface should be relatively sharp (i.e. L small) because it is an interface between two liquids. The situation is sketched in Fig. 8. We can see from Fig. 9 that especially in the case of small L there are reflection coefficients in the order of some percents that eventually should be taken into account for seismic interpretation.

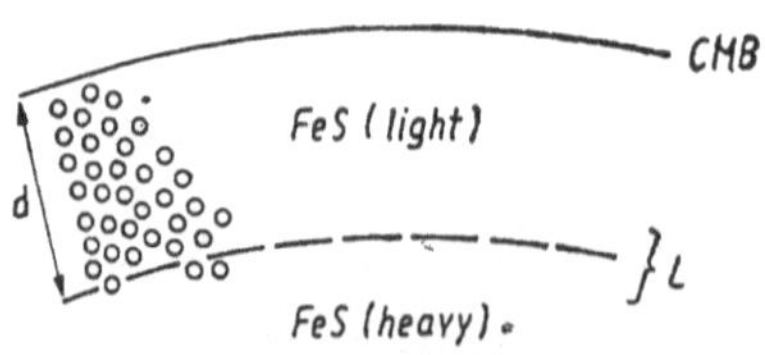

Fig. 8. Thickness and roughness of the stably stratified layer.

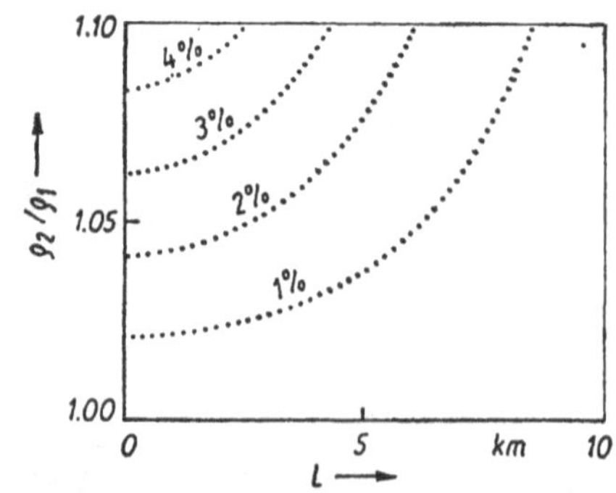

Fig. 9. Reflection coefficients for the stably stratified layer in dependence of density contrast and roughness.

Scattering of seismic waves

On the base of our model, data from the seismic station MOX in the GDR have been reevaluated under the assumption that the scattering inhomogenities for seismic waves are at the base of the stably stratified layer, that means about 10—15 km below the CMB (see also STILLER et al., 1985). The calculated minimum travel time curve deviates only very little from the minimum travel time curve for scattering directly at CMB (Model B1 of JORDAN and ANDERSON, 1974). Fig. 10 shows that our calculated minimum travel time curve envelopes all observational data. In this way there is also a good conformity with our model. Concluding, one can say that our model of Earth's core evolution seems to be favourable because it is related to modern thinking of planetogony, new results of high-pressure physics, and possible observations with the help of seismic waves.

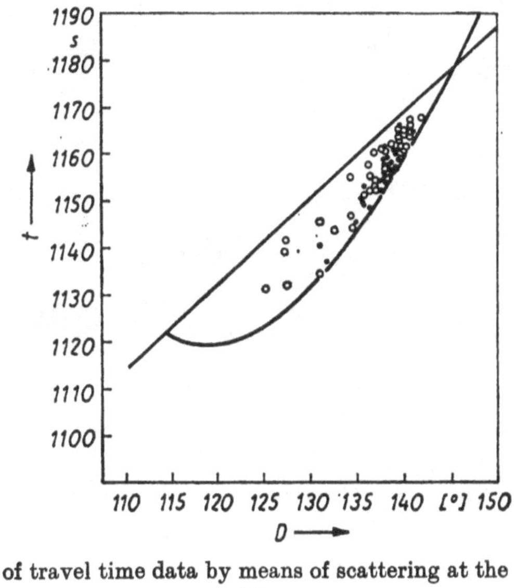

Fig. 10. Interpretation of travel time data by means of scattering at the base of the stably stratified layer.

References

AHRENS, T. J.: Equations of state of iron sulfide and constraints on the sulfur content of the Earth. J. Geophys. Res. 84 (1979) B3, 985—998.

ELSASSER, W. M.: Early history of the Earth. In: J. GEISS (Editor): Earth Sciences and Meteoritics. Wiley, New York (1963) 1.

FRANCK, S.: Ascending droplets in the Earth's core. Phys. Earth Planet. Inter. 27 (1982) 249—254.

JORDAN, T. H.; ANDERSON, D. L.: Earth structure from free oscillations and travel times. Geophys. J. 36 (1974) 411—420.

LESS, A. C.; BUKOWINSKI, M. S. T.; JEANLOZ, R.: Reflection Properties of Phase Transition and Compositional Change Models of the 670 km Discontinuity. J. Geophys. Res. 88 (1983) B10, 8145—8159.

Loper, D. E.; Roberts, P. H.: In: The fluid mechanics of astrophysics and geophysics (ed. A. M. Soward), vol. 2 (Stellar and planetary magnetism). New York, London and Paris: Gordon & Breach Science Publications, 1982.

Safronov, V. S.: Evolution of the Preplanetary Cloud and the Formation of the Earth and Planets (in Russian). Nauka, Moscow 1969.

Schmidt, O. Yu.: A Theory of the Origin of the Earth. Four Lectures., transl. by G. H. Manna, Lawrence and Wishart, London 1959.

Stiller, H.; Franck, S.; Kowalle, G.: On the formation of a stably stratified layer near core-mantle-boundary. submitted to Journal of Geodynamics (1985).

Author's Index

Bakiev, Muz. K. (3)
Bayuk, E. I. (6)
Bayuk, I. O. (1)
Balakishiev, S. A. (4)
Czechowski, Z. (14)
Danukalov, K. N. (1)
Dyaur, N. I. (6)
Efimova, G. A. (1)
Egerer, F. (11)
Emirov, S. N. (5)
Franck, S. (16)
Freund, D. (16)
Fojtek, A. (9)
Gálos, M. (12)
Gasanov, A. B. (4)
Ibragimov, A. K. (3)
Kalinin, V. A. (1)
Kapička, A. (8)
Kopf, M. (16)
Korchin, V. A. (2)
Kraft, A. (16)
Krešl, M. (8)
Kropáček, V. (8)
Laštovičková, M. (8)
Lebedev, T. S. (2)
Leliwa-Kopystyński, J. (13)
Levitova, E. M. (6)
Martinec, P. (10)
Novik, G. Y. (7)
Parhomenko, E. I. (1)
Podroužková, Z. (8)
Pralat, A. (15)
Rodkin, M. V. (1)
Savenko, B. Y. (2)
Seifert, W. (16)
Seipold, U. (16)
Shepel, S. I. (2)
Silbersmit, M. G. (7)
Stiller, H. (16)
Švec, J. (9)
Tomashevskaya, I. S. (1)
Ullmann, W. (17)
Zavorikina, T. K. (7)
Zhukov, I. V. (1)
Ziemann, M. (16)

Authors' Addresses

(1) Institute of Physics of the Farth, Academy of Sciences of the USSR, Bolshaya Gruzinskaya 10, 123810 Moscow D-242, USSR
(2) Institute of Geophysics, Academy of Sciences of the Ukrainian SSR, Palladina 32, 252068 Kiev — 68, USSR
(3) Institute of Seismology, Uzbek Academy of Sciences, Khurshida ul. 3, 700128 Tashkent, USSR
(4) Geophysical Institute, Academy of Sciences of the Azerbaijan SSR, Narimanova 29a, 370143 Baku, USSR
(5) Institute of problems of Geothermics, Daghestan Branch of USSR Academy of Sciences, Ul. Bakinskich Kommissarow 75, 367002 Makhachkala, USSR

(6) Institute of Geology and Prospecting, Fersmana 50, 117312 Moscow, USSR
(7) Institute of Mines, Leninprospect 6, Moscow, USSR
(8) Geophysical Institute, Czechoslovak Academy of Sciences, Boĉni II 1401, 141 31 Praque 4, CSSR
(9) Technical University Ostrava, Vítězného února 3, 708 33 Ostrava, CSSR
(10) Coal Research Institute, Pikartská ul., 716 07 Ostrava-Radvanice, CSSR
(11) Technical University for Heavy Industry, Department of Mineralogy and Petrology, 3515 Miskolc-Egyetemvaros, Hungary
(12) Technical University Budapest, Department of Mineralogy and Geology, Stoczek utca 2, 1521 Budapest XI, Hungary
(13) Institute of Geophysics, Warsaw University, ul. Pasteura 7, 02-093 Warszawa, Poland
(14) Institute of Geophysics, Polish Academy of Sciences, ul. Pasteura 3, 00-973 Warszawa, Poland
(15) University of Wroclaw, Place Universytecki 1, Wroclaw, Poland
(16) Institute of High Pressure Research, Academy of Sciences of the GDR, Telegrafenberg, 1561 Potsdam, GDR
(17) Central Institute of Physics of the Earth, Academy of Sciences of the GDR, Burgweg 11, 6900 Jena, GDR

FSC
www.fsc.org
MIX
Papier | Fördert gute Waldnutzung
FSC® C083411

Zeitfracht Medien GmbH
Ferdinand-Jühlke-Straße 7
99095 Erfurt, Deutschland
produktsicherheit@kolibri360.de